SOLAR SYSTEM

A Visual Exploration of the Planets, Moons, and Other Heavenly Bodies that Orbit Our Sun

WRITTEN BY MARCUS CHOWN

Published in partnership with Touch Press, creators of *The Elements* and *Solar System for iPad*; Faber and Faber; and Planetary Visions

SOLAR S

YSTEM

WRITTEN BY MARCUS CHOWN

Published in partnership with Touch Press, creators of *The Elements* and *Solar System for iPad*; Faber and Faber; and Planetary Visions

A Visual Exploration of the Planets, Moons, and Other Heavenly Bodies that Orbit Our Sun

BLACK DOG & LEVENTHAL PUBLISHERS
NEW YORK

C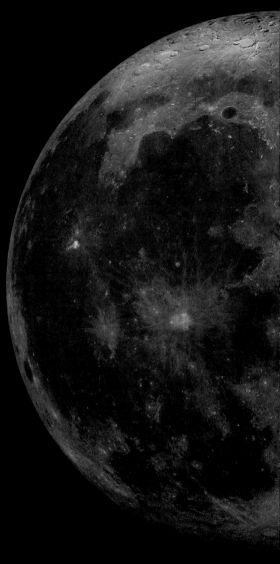ntents

The Data and Images Explained

MOST OF THE PICTURES in this book have been taken by space probes sent out to explore our neighboring planets over the last thirty years. Individual pictures, selected from the many thousands that are available, range in scale from whole-disc images taken by ground-based or space telescopes, to microscopic views of rock structure from rover on-board cameras. As well as visible-light images, the full spectrum is shown from x-rays, through ultraviolet and infrared to radio emissions, reflecting the range of sensors used to probe planetary surfaces, atmospheres and magnetic fields.

The planet and moon maps are compiled from many images, sometimes many hundreds, taken by spacecraft as they orbited a planet, or simply flew past it. Each image is geometrically adjusted to show part of the planet's surface and then blended with others, correcting for illumination variation, to build up the global map.

The inner planets have each been visited by multiple space probes: Mariner 10 and MESSENGER to Mercury; Venera, Magellan, and Venus Express to Venus; a whole fleet of satellites and landers to Mars.

Multiple missions have also made it to the large outer planets: Galileo orbited Jupiter; Cassini is still orbiting Saturn, having dropped a lander onto its moon Titan; before that the Voyager mission flew spacecraft past all four gas giants, with Voyager 2 being the only spacecraft to have visited Uranus and Neptune.

Today the New Horizons spacecraft is on its way to Pluto and the Kuiper Belt. Robot probes have even visited asteroids and brought back rock samples, chased comets and returned their dust.

Some planets are better mapped than the Earth, where more than 70% of the solid surface is hidden by deep water. Other bodies are only partly covered by images due to the fleeting nature of a flyby encounter. There you will see blank areas on the maps–not quite "Here be Dragons," but definitely "terra incognita." The most distant objects, including newly-discovered dwarf planets, are no more than a few pixels across in images from the world's most powerful telescopes–tantalising glimpses of new lands yet to be explored.

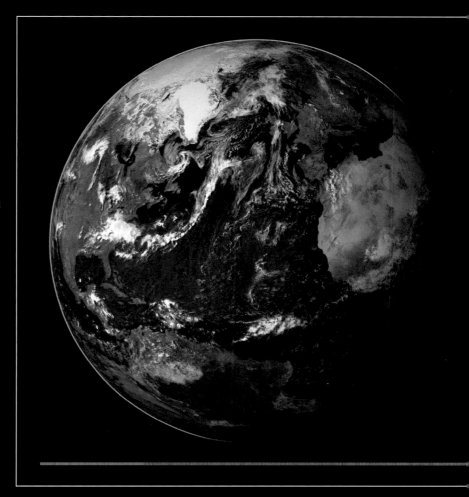

PLANET AND MOON GLOBES ▲

These computer-graphic views show each planet, moon or asteroid from a similar viewpoint and with similar illumination, so their appearance can be directly compared. These views are generated from global maps of each body based on the best available space probe imagery. Each body is shown as close as possible to its natural appearance, so Earth, Venus, and Saturn's moon Titan are shown with their characteristic cloud cover. The planets are shown with their correct axial tilt, with a rotation angle (longitude) chosen to show some of their key surface features.

▶ SOLAR SYSTEM MAPS

The double-page 3D maps of the Solar System are accurate computer graphic simulations with planet orbits plotted to true scale. For clarity, the size of the planets and moons in most of the maps has been exaggerated by 500 times, and moon orbits are shown at 50 times their correct size. The position of each body is correct for January 1, 2011. The stars shown in the background are accurate too, drawn from a photographic all-sky survey, with features such as the Milky Way and the Magellanic Clouds appearing in several of the views.

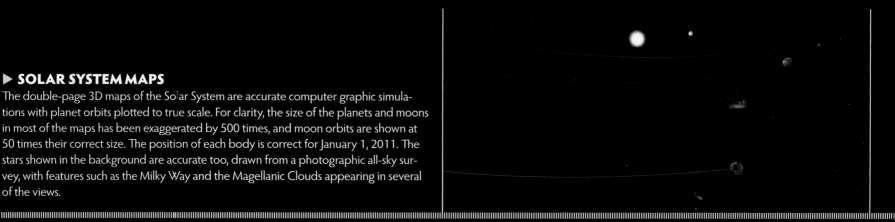

Earth

IT IS BIG. It is round. And we all depend on it. When considering a world as familiar as Earth, it is hard to say anything new. But our planet is deeply mysterious. It is the only world with surface water. It is the only world with plate tectonics, an ozone layer–and life. Why is the Earth so special? It must be related to its distance from the Sun–smack bang in the Goldilocks zone, not too hot, not too cold–to its mass and composition, its giant moon, which stabilizes climate. Whereas the most complex thing on other planets is weather, on Earth complexity has increased remorselessly–from bacteria to multi-cellular life, human society, civilization, and technology. If you know why here and nowhere else, there is a Nobel Prize waiting for you. In the meantime, consider the ways the Earth is unique.

ORBITAL DATA
Distance from Sun 147 to 152 million km / 0.98 to 1.02 AU
Orbital Period (Year) 365.26 Earth days
Length of Day 23.935 Earth hours
Orbital Speed 30.3 to 29.3 km/s
Orbital Eccentricity 0.0167
Orbital Incination 0°
Axial Tilt 23.44°

- Mercury
- Venus
- Earth
- Mars

PHYSICAL DATA
Diameter 12,756 km
Mass 5,970 billion billion metric tons
Volume 1,080,000 million km³
Gravity 1 x Earth
Escape Velocity 11.18 km/s
Surface Temperature 204° to 331°K / −69° to 58°C
Mean Density 5.515 g/cm³

The moon

ATMOSPHERIC COMPOSITION
Nitrogen 78.084%
Oxygen 20.946%
Argon 0.9340%
Water vapor 0.1000%
Carbon dioxide 0.039%
Neon 0.001818%
Helium 0.000524%
Methane 0.000179%
Krypton 0.000114%
Hydrogen 0.000055%
Nitrous oxide 0.00003%
Carbon monoxide 0.00001%

EARTH 55

Labels on cross-section:
- Nitrogen/oxygen atmosphere
- Seas and oceans
- Rocky crust
- Silicate upper mantle
- Solid iron–nickel inner core
- Liquid iron–nickel outer core
- Silicate lower mantle

PLANET CROSS-SECTIONS ▲
The cross-section views show the interior structure of each planet, as far as we can ascertain it, from atmosphere or crust to core. Internal structure is inferred from the planet's mass and size, and the laws of physics.

▼ PLANET AND MOON MAPS
The planet and moon maps show the entire surface of each body, with Earth, Venus, and Titan shown without clouds. The maps are in the equal-area Mollweide projection, which keeps the relative size of features correct, although there is distortion of shape toward the edges and poles. This is just one way of representing a planet's spherical surface on a flat piece of paper–like the skin of an orange peeled off in one piece and flattened out.

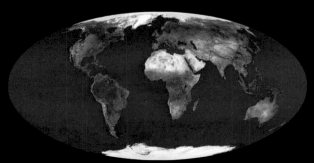

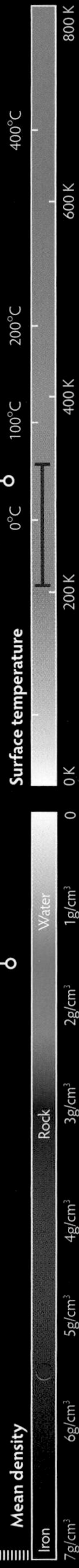

◀ PLANET AND MOON DATA
The "vital statistics" of each body are grouped into Orbital Data, describing where the body is and how it moves, and Physical Data, describing the size, mass and other physical properties of the body itself.

Two important properties are shown visually on scale bars at the side of the page; **Surface Temperature** follows a pattern as we move away from the Sun's warmth, with the odd exception such as Venus, whose temperature is elevated by the greenhouse effect of its thick atmosphere. Only Earth lies comfortably in the temperature range 0-100°C where liquid water, so important to life, can exist on the surface.

Mean Density gives us a clue about what the planet or moon may be made of, with small, solid Mercury having a density approaching that of iron, and giant, gassy Saturn being less dense than water.

ORBIT MAPS
The orbit map shows the shape of the orbit of each planet, moon or asteroid in relation to its neighbors and parent body. The orbits are to true scale and the positions of each body are correct for January 1, 2012.

SCALE MAPS
The scale map shows the size of each planet, moon or asteroid in comparison to an object that is (we hope!) more familiar. The scale objects range in size from the Earth down to a human being. (The waving human outline we've used to scale the particles of Saturn's rings is taken from the plaque on the side of the Pioneer 10 space probe, drawn by Linda Salzman Sagan, wife of astronomer Carl Sagan.)

Surface temperature scale (right margin): 800 K, 400°C / 600 K, 200°C / 400 K, 100°C / 200 K, 0°C / 0 K

Mean density scale (right margin): 0 / Water 1g/cm³ / 2g/cm³ / Rock 3g/cm³ / 4g/cm³ / 5g/cm³ / 6g/cm³ / Iron 7g/cm³

Solar System

Saturn ▶

Mercury ▼ ◀ Venus

Mars ▲ ▲ Earth

◀ Jupiter

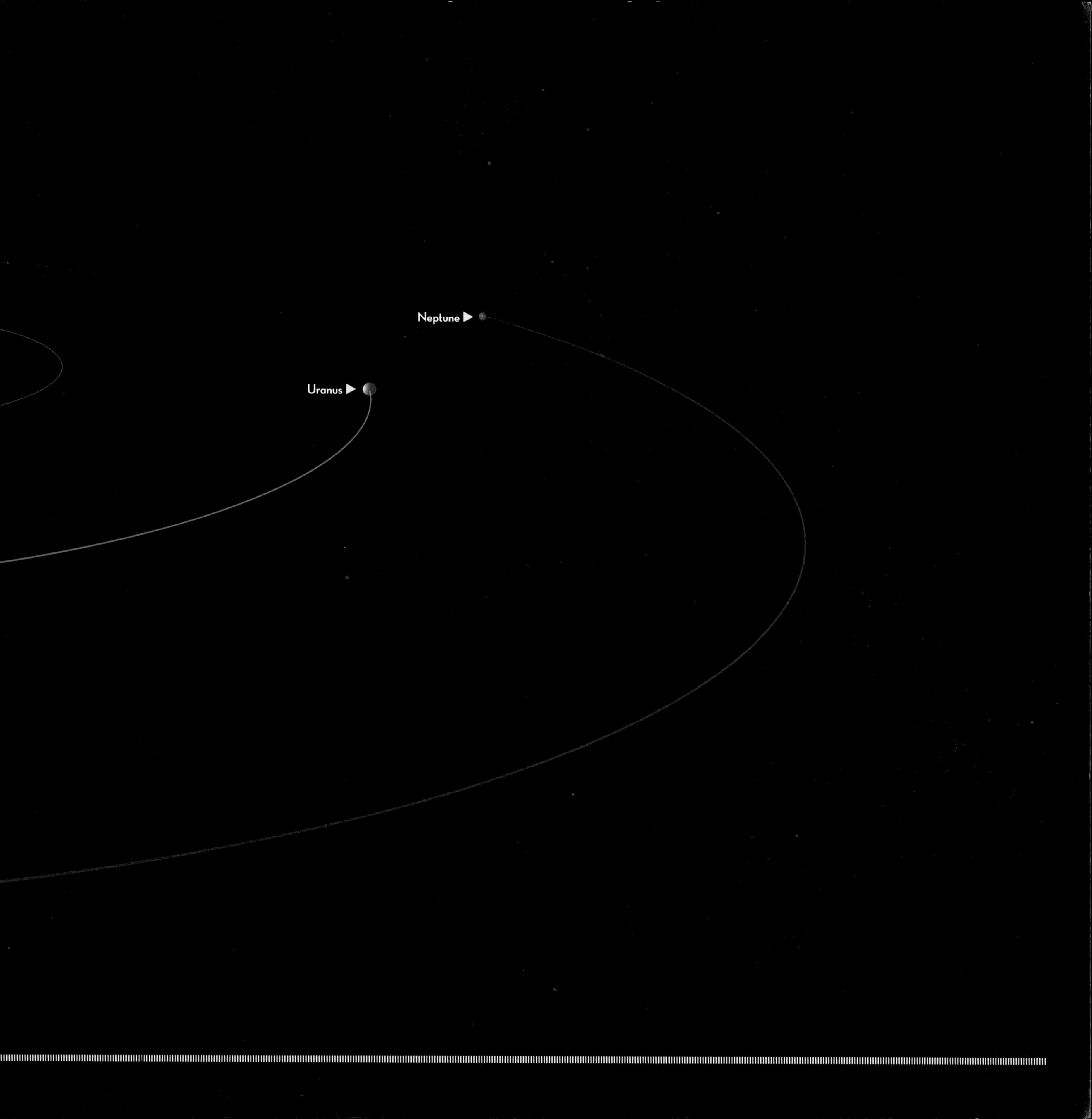
Neptune ▶
Uranus ▶

Solar System

FOR MOST PEOPLE, life on Earth is hard. For the lucky ones it is merely hectic. No wonder we get caught up in our daily lives. We look down rather than up. We ignore the fact that we live on a tiny speck of rock suspended in an unutterable vastness of empty space. But above the thin skin of the atmosphere are other worlds. Worlds where 100-year-old hurricanes are raging, ice volcanoes are erupting, gargantuan

lightning bolts are leaping between cloud tops and moons. These things have been happening for billions of years, but it is only now that we can see them in glorious close-up. We are extraordinarily privileged to be alive at the very beginning of the age of planetary exploration. Welcome to the Sun and planets, the moons and comets and chunks of assorted rubble that make up the Solar System.

NUMBER OF PLANETS
8 (Mercury, Venus, Earth, Mars, Jupiter, Saturn, Uranus, and Neptune)

NUMBER OF DWARF PLANETS
5 (Ceres, Pluto, Eris, Haumea, and Makemake)

PLANETARY MOONS
162

DIAMETER
64,000,000 million km / 427,813 AU (outer limit of the Oort Cloud)

What is the Solar System?

THE SOLAR SYSTEM is the collection of bodies under the gravitational influence of the Sun, essentially the Sun plus a tiny amount of builders' rubble left over from its birth 4.55 billion years ago. Although the Sun contains 99.8 percent of the Solar System's mass, the builders' rubble contains most that is of interest. Then again, we would say that, since included in the rubble is the Earth.

The main components of the Solar System in order of increasing distance from the Sun are four rocky, or terrestrial, planets–Mercury, Venus, Earth, and Mars–and four "gas giants"–Jupiter, Saturn, Uranus, and Neptune. Between the two groups orbits a swarm of rocky rubble known as the Asteroid Belt, and, beyond the gas giants, a swarm of icy rubble called the Kuiper Belt. Far, far beyond everything else is the "Oort Cloud," containing maybe a trillion icy comets.

Imagine a CD in a giant swarm of bees. This is how it is for the planets, asteroids, and so on within the Oort Cloud.

As for scale, if the Sun were the size of a peppercorn, the Earth would be 10 centimeters away, and Eris, the biggest Kuiper Belt object, 10 meters away. The Oort Cloud, however, extending halfway to the next star, would be 10 kilometers in radius. This marks the extent of the Sun's gravitational influence and the edge of the Solar System.

So we know what it is, but where is the Solar System?

Mercury

Venus

Earth

Mars

Jupiter

Saturn

Uranus

Neptune

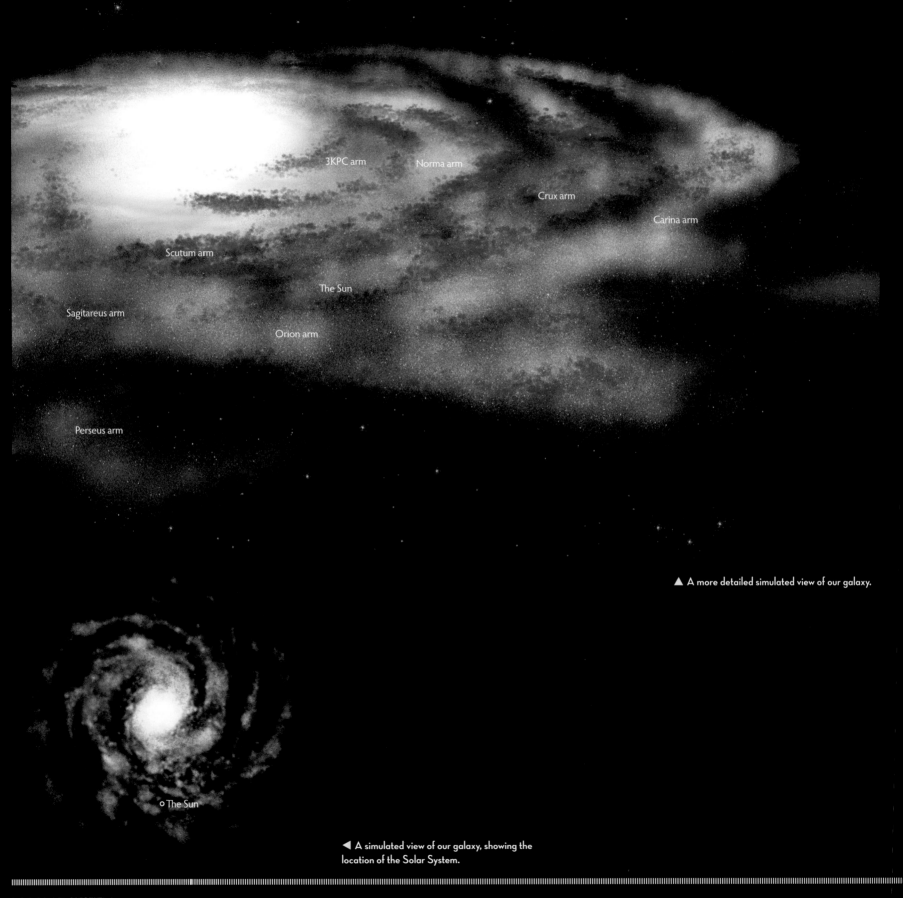

3KPC arm

Norma arm

Crux arm

Carina arm

Scutum arm

The Sun

Sagitareus arm

Orion arm

Perseus arm

▲ A more detailed simulated view of our galaxy.

○ The Sun

◀ A simulated view of our galaxy, showing the
location of the Solar System.

Where is the Solar System?

THE SUN IS one of about 100 billion stars in a spiral galaxy called the Milky Way, a great whirlpool of stars turning ponderously in space. Edge-on, the Milky Way would look like two fried eggs back-to-back. This whole system of stars is believed to be embedded in a giant spherical halo of invisible–"dark"–matter.

The Sun orbits the center of the Galaxy at a distance of about 26,000 light-years in a "spiral arm" that snakes out from the central bulge of stars. This arm–actually a spur of the Perseus spiral arm–is roughly halfway to the outer edge. At this distance, the Sun goes around the center of the Galaxy once in about 220 million years. As it orbits, it also oscillates up and down through the plane of the Galaxy. Encounters with "Giant Molecular Clouds" of gas may stir up the Oort Cloud, sending comets sunward–to strike the Earth and trigger mass extinctions?

The Milky Way is one of about 100 billion galaxies in the Observable Universe, a bubble about 84 billion light years across and centered on the Earth. It contains all galaxies whose light has had time to reach us since the universe's birth 13.7 billion years ago. Beyond its edge, or "light horizon," are other galaxies currently invisible to us. Conceivably, the universe may march on forever. Like the universe, the Solar System has not existed forever.

▲ The Milky Way is visible in the northern hemisphere, but it truly shines south of the equator. This is because Earth's south pole is tipped toward the center of the galaxy, where most of the stars are concentrated, whereas the north pole points out toward the edge of the galaxy.

Where did it come from?

IN THE BEGINNING, about 4.55 billion years ago, there was a Giant Molecular Cloud. Such clouds exist in our galaxy today–clumps of gas and dust, cold and dark, like pools of ink spilled across the stars. The primordial cloud would have hung there for ages, doing nothing, but for the molecules it was made of. The light they gave out escaped the cloud, taking heat with it. Deprived of its internal heat, the cloud could not prevent gravity from crushing it. It began to shrink.

When spinning things shrink–and the cloud shared the Galaxy's rotation–they spin faster, like an ice-skater pulling in her arms. Simultaneously, the cloud fragmented into smaller globules, one of which would become our solar system.

The globule grew hotter as it continued to shrink until, at the center, it was hot enough to ignite energy-generating nuclear reactions. The Sun was born. Around the newborn Sun swirled a debris disk of leftover gas and dust. It was flat because a rotating gas cloud shrinks more easily between its poles than around its equator, where gravity is hindered by the "centrifugal" tendency of matter to fly outward.

In the debris disk, dust particles hit each other and stuck, building ever-bigger bodies that eventually aggregated into planets. Computer simulations commonly show about 10 Earth-mass bodies forming. Most are catapulted out of the Solar System by close encounters with embryonic giant planets–Earth's brothers and sisters, lost forever in the ocean of interstellar space.

1.

▼ Stars are born when immense clouds of gas and dust are compressed into dense structures, such as these dust pillars in the Carina Nebula. The pillars are being sculpted by violent stellar winds and powerful radiation from nearby massive stars.

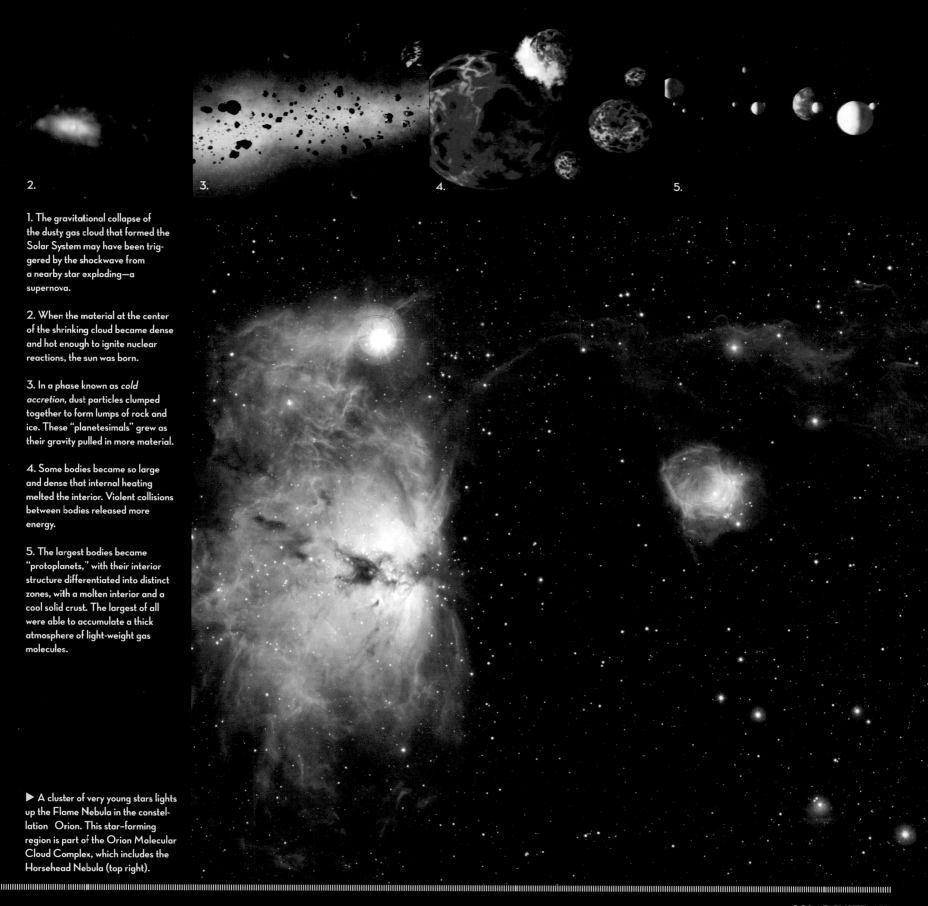

2.

3.

4.

5.

1. The gravitational collapse of the dusty gas cloud that formed the Solar System may have been triggered by the shockwave from a nearby star exploding—a supernova.

2. When the material at the center of the shrinking cloud became dense and hot enough to ignite nuclear reactions, the sun was born.

3. In a phase known as *cold accretion*, dust particles clumped together to form lumps of rock and ice. These "planetesimals" grew as their gravity pulled in more material.

4. Some bodies became so large and dense that internal heating melted the interior. Violent collisions between bodies released more energy.

5. The largest bodies became "protoplanets," with their interior structure differentiated into distinct zones, with a molten interior and a cool solid crust. The largest of all were able to accumulate a thick atmosphere of light-weight gas molecules.

▶ A cluster of very young stars lights up the Flame Nebula in the constellation Orion. This star–forming region is part of the Orion Molecular Cloud Complex, which includes the Horsehead Nebula (top right).

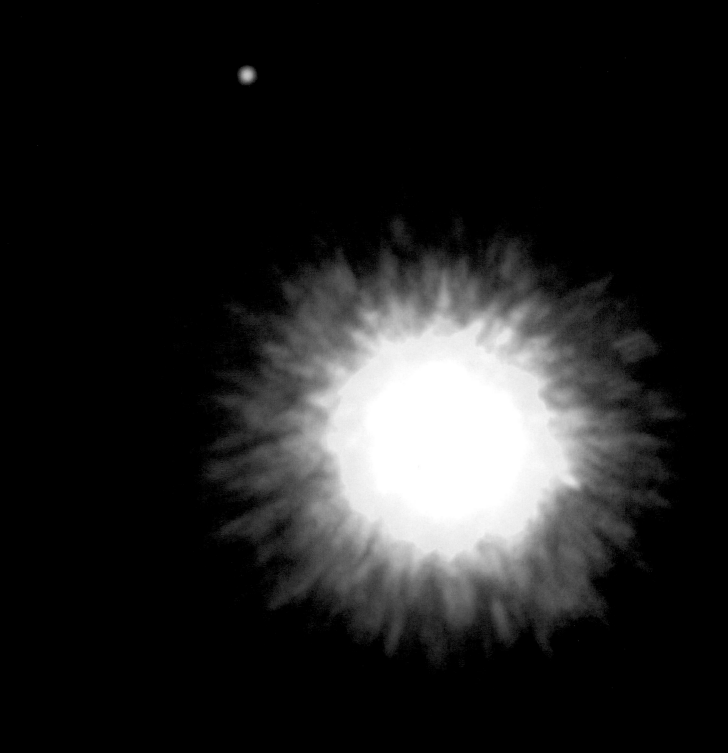

▲ The first picture of an alien planet orbiting around a sun-like star. The planet is about eight times as massive as Jupiter, and orbits its parent at a distance ten times greater than Neptune's orbit of the Sun.

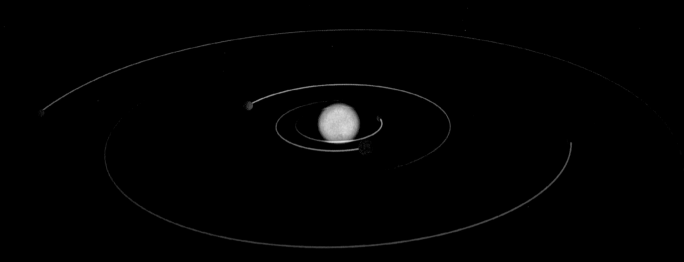

Is the Solar System unique?

BEFORE 1995 WE had no idea whether planets like ours orbited other stars. Since then, huge numbers of "extrasolar" planets have been discovered. It seems that about one in ten nearby stars has a planet and maybe one in three a protoplanetary disk. Extrapolating to the whole Milky Way, we could be talking about tens of billions of planetary systems. The closest known planetary system is around Epsilon Eridani, 10.5 light-years away.

The huge surprise has been how very different other planetary systems are from our own. Some even have gas giant planets orbiting closer to their star than Mercury is to the Sun. Such "hot Jupiters" could not possibly have formed in situ since the gas necessary to create their mantles would have been too hot for their gravity to trap. Friction between an embryonic giant planet and a primordial debris disk may have caused a planet to migrate toward its star.

It is difficult to know whether the planetary systems discovered so far are typical since the techniques used to find them tend to reveal only very massive planets. Those techniques involve observing the wobble of a star as it is tugged by the gravity of an orbiting planet, or observing the dip in the brightness of a star when a massive planet passes in front of, or transits, the star. We are getting close to being able to detect an Earth-mass world around a Sun-like star. The aim is to find another Eden.

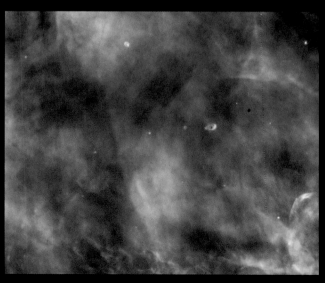

▲ Of these five young stars in the Orion Nebula, four have discs of gas and dust left over from their formation. These protoplanetary discs (or "proplyds") may evolve into planetary systems around the stars.

Space

IT IS EASY to overlook the biggest component of the Solar System, one that utterly overwhelms everything else: space.

In interplanetary space there are on average only ten atoms and molecules per cubic centimeter. For comparison, the same volume of air at sea level contains about 30 billion billion, while a good terrestrial vacuum has about 100,000 molecules.

Since the interplanetary vacuum is essentially empty, it is true that, in space, no one can hear you scream. Sound, after all, requires a medium to vibrate in. Light requires no medium. However, laser beams in space are invisible because they can be seen only if there is dust to scatter the light from the beam into your eye (apologies to Star Wars fans).

An astronaut in space faces many hazards. They must carry a supply of air since there is no air to breathe. Their suit must be heated since the vacuum, with so few molecules, is mind-cringingly cold. But their suit must also have the capacity to be cooled since there is no air to carry away surplus heat, and an astronaut, caught in direct sunlight, risks overheating. In addition, the astronaut's suit must be pressurized because they,

like you and me, usually live with the weight of 50 kilometers of air weighing down on them—equivalent to a kilogram on every square centimeter of their body—and, without this, the astronaut's blood will boil. As if all this were not bad enough, there is the constant danger of deadly radiation from solar flares. Ever get the feeling that humans were not made for space?

▶ Astronaut Bruce McCandless floats untethered in space, testing the Manned Maneuvering Unit during Shuttle Mission 41B in 1984.

Life in the Solar System

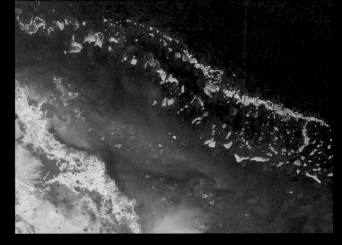

WATER IS a prerequisite for life. We may be wrong about that but we have only one example of life to go on: our own. And we need water. Where water can exist in the Solar System is therefore crucial in knowing where to look for extraterrestrial biology.

For liquid water to exist it must be hot enough that it does not freeze and cold enough that it does not boil. Since temperature falls with increasing distance from the Sun, there is a Goldilocks region known as the Habitable Zone in which a planet can orbit and have surface water. It extends from just inside the Earth's orbit almost out to Mars. The zone is broadened a little by the presence of greenhouse gases, such as water vapor on Earth, which trap heat.

Sometimes this band is called the "classical" Habitable Zone. This is because recent discoveries have blurred the concept. For instance, differences in gravity across a body, or tidal forces, can stretch and squeeze it, heating the interior. Jupiter's gravity is doing this to both Io, which is hot enough to sport active volcanoes, and Europa, which is believed to have an ocean beneath its icy surface. Both moons are further from the Sun than the classical Habitable Zone.

Most excitingly, Saturn's moon Titan appears to have oceans and rivers, rain and snow, composed not of water but liquid methane and ethane. This raises the tantalizing possibility that a liquid other than water might provide a medium in which life's chemicals interact, creating a truly alien biology.

▲ The Great Barrier Reef is the largest single structure made of living organisms—a collection of coral reefs and islands stretching more than 2,600 km along the northeastern coast of Australia. Coral reefs are one of the richest habitats for life on Earth, providing a home for 25% of all marine species.

An alien artifact?

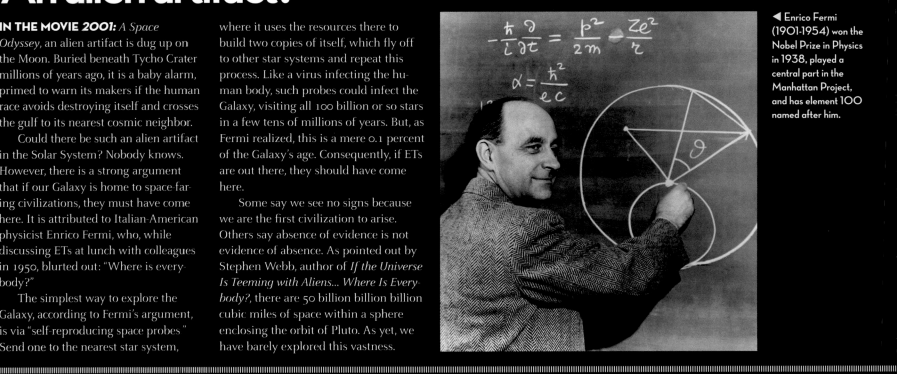

IN THE MOVIE 2001: *A Space Odyssey*, an alien artifact is dug up on the Moon. Buried beneath Tycho Crater millions of years ago, it is a baby alarm, primed to warn its makers if the human race avoids destroying itself and crosses the gulf to its nearest cosmic neighbor.

Could there be such an alien artifact in the Solar System? Nobody knows. However, there is a strong argument that if our Galaxy is home to space-faring civilizations, they must have come here. It is attributed to Italian-American physicist Enrico Fermi, who, while discussing ETs at lunch with colleagues in 1950, blurted out: "Where is everybody?"

The simplest way to explore the Galaxy, according to Fermi's argument, is via "self-reproducing space probes." Send one to the nearest star system, where it uses the resources there to build two copies of itself, which fly off to other star systems and repeat this process. Like a virus infecting the human body, such probes could infect the Galaxy, visiting all 100 billion or so stars in a few tens of millions of years. But, as Fermi realized, this is a mere 0.1 percent of the Galaxy's age. Consequently, if ETs are out there, they should have come here.

Some say we see no signs because we are the first civilization to arise. Others say absence of evidence is not evidence of absence. As pointed out by Stephen Webb, author of *If the Universe Is Teeming with Aliens... Where Is Everybody?*, there are 50 billion billion billion cubic miles of space within a sphere enclosing the orbit of Pluto. As yet, we have barely explored this vastness.

◄ Enrico Fermi (1901-1954) won the Nobel Prize in Physics in 1938, played a central part in the Manhattan Project, and has element 100 named after him.

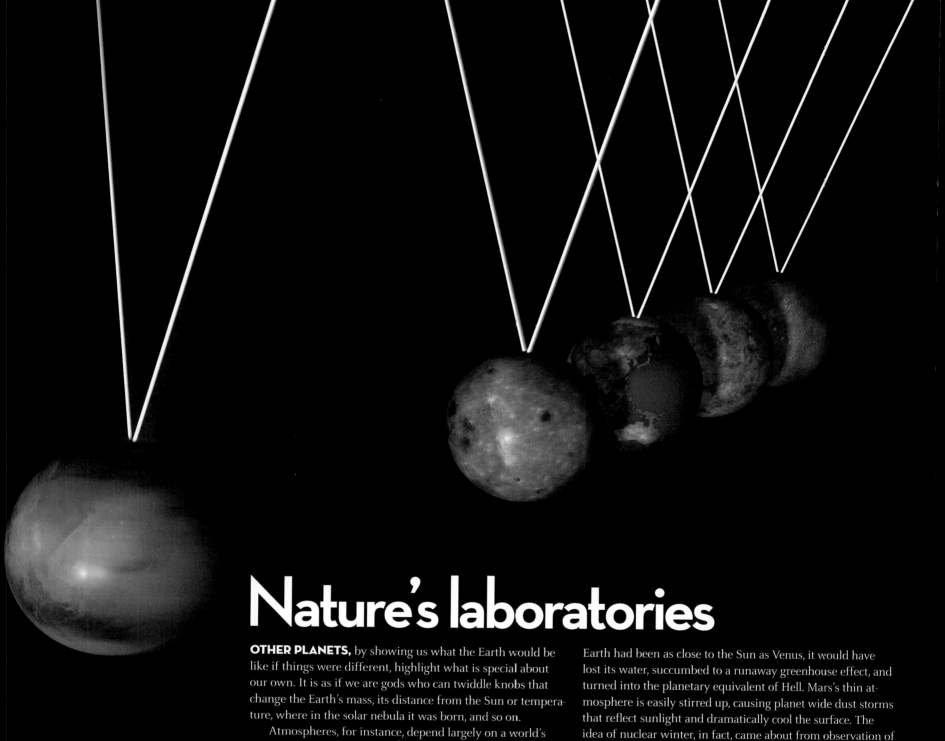

Nature's laboratories

OTHER PLANETS, by showing us what the Earth would be like if things were different, highlight what is special about our own. It is as if we are gods who can twiddle knobs that change the Earth's mass, its distance from the Sun or temperature, where in the solar nebula it was born, and so on.

Atmospheres, for instance, depend largely on a world's mass and distance from the Sun. Close in, where it is hot, gas molecules are flying about fast like angry bees, so it requires the strong gravity of a massive planet to hold them. Mercury is too small. Venus and Earth are massive enough, so they do hold onto an atmosphere. Far from the Sun, however, where it is cold and gas molecules are moving sluggishly, even a small world like Saturn's moon Titan can keep a thick atmosphere.

There are also more subtle and complex effects. If the Earth had been as close to the Sun as Venus, it would have lost its water, succumbed to a runaway greenhouse effect, and turned into the planetary equivalent of Hell. Mars's thin atmosphere is easily stirred up, causing planet wide dust storms that reflect sunlight and dramatically cool the surface. The idea of nuclear winter, in fact, came about from observation of Mars. Both worlds, Mars and Venus, serve as warnings to the human race.

The incredible thing is that the planets continue to surprise us with, for instance, the recent discovery of water on Saturn's tiny moon Enceladus. Yes, planets are nature's laboratories, but the interplay of different forces is so complex that we often cannot guess the result. We have to go there and see for ourselves.

The Inner Solar System

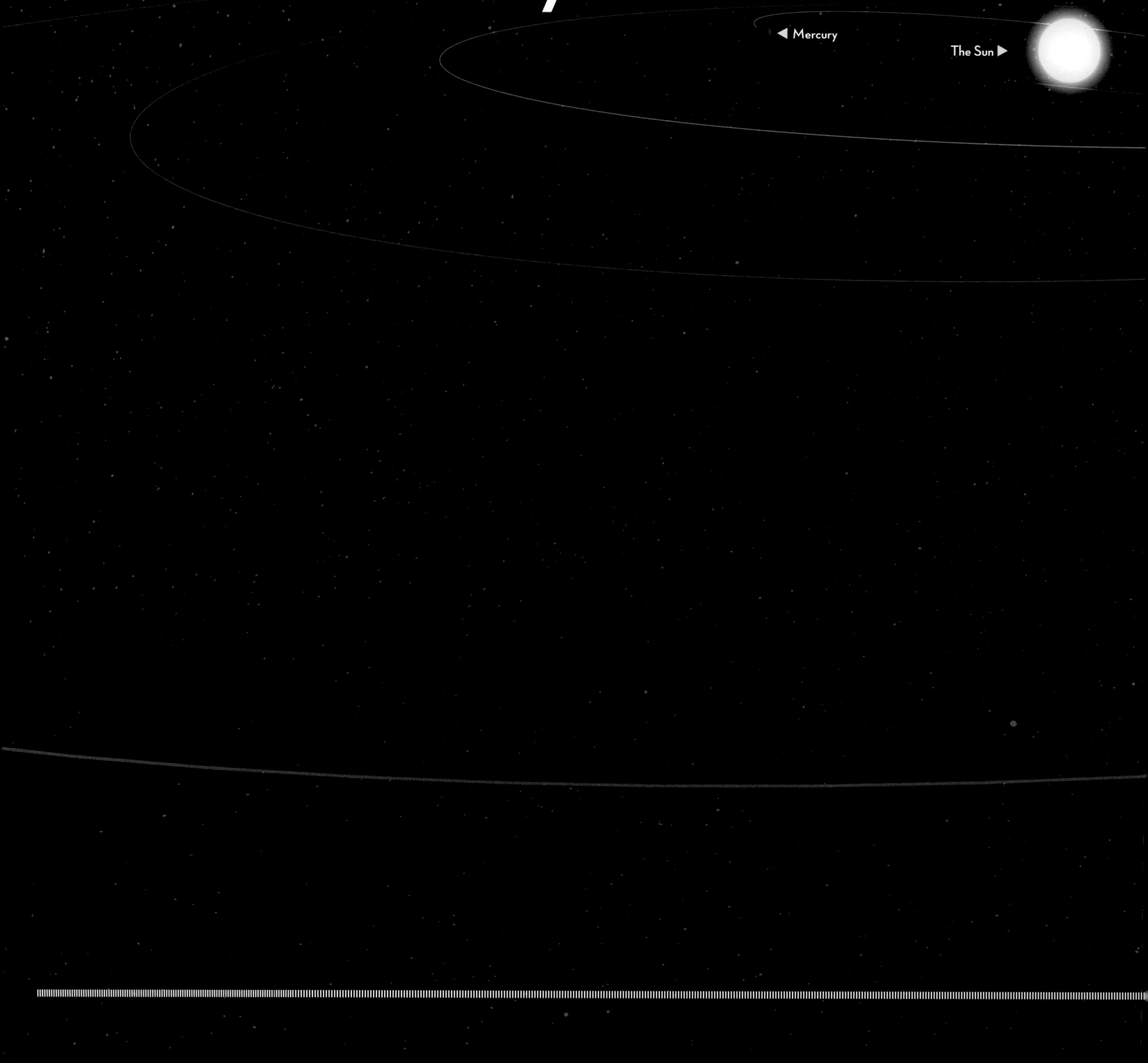

◀ Mercury

The Sun ▶

Venus ▶

◀ The Moon

◀ Earth

▼ Mars

Phobos ▶   ◀ Deimos

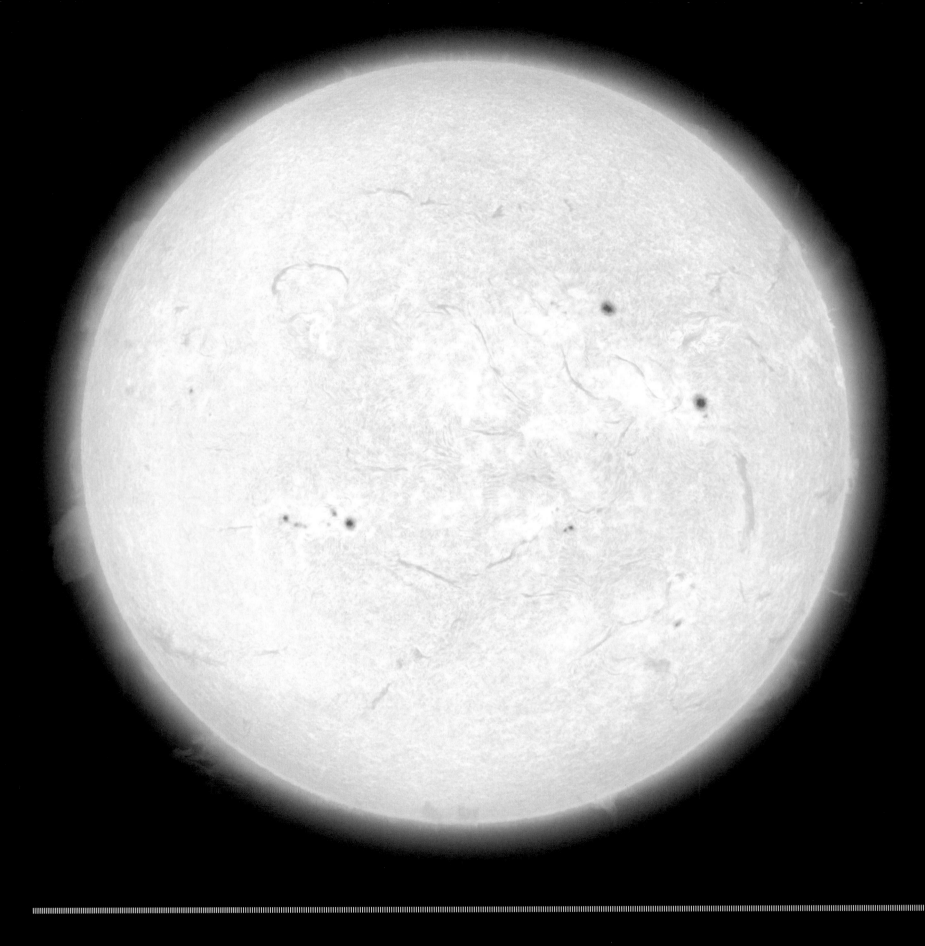

Sun

THE SUN is the nearest star. It is the only star close enough to appear as a disk rather than a mere pinprick of light. Essentially, the Sun is the Solar System. It accounts for 99.8 percent of the mass and is so big a million Earths would fit inside. It has been pumping out light and heat, more or less constantly, since the birth of the Earth about 4.55 billion years ago. This highlights a fundamental difference between a star and a planet. A star generates its own heat and light, whereas a planet, formed in the debris disk around a star, shines (mostly) from reflected light. Like all stars, the Sun is a giant ball of gas, held together and squeezed by its own gravity until it is superhot. But what is that gas? In short, what is the Sun made of?

ORBITAL DATA
LENGTH OF DAY 27 Earth days
AXIAL TILT 7.25°

PHYSICAL DATA
Diameter 1,391,900 km / 109 x Earth
Mass 1.99 billion billion billion metric tons/ 333,333 x Earth
Volume 1.41 million million million km³ / 1.3 million x Earth
Gravity 27.963 x Earth
Escape Velocity 617 km/s
Surface Temperature 5,780°K / 5,507°C
Mean Density 1.41 g/cm³

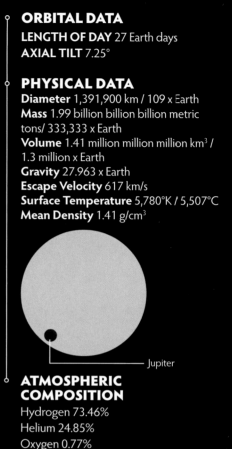

Jupiter

ATMOSPHERIC COMPOSITION
Hydrogen 73.46%
Helium 24.85%
Oxygen 0.77%
Carbon 0.29%
Iron 0.16%
Sulfur 0.12%
Neon 0.12%
Nitrogen 0.09%
Silicon 0.07%
Magnesium 0.05%

— Chromosphere

— Photosphere

— Convective zone

— Radiative zone

— Rocky core

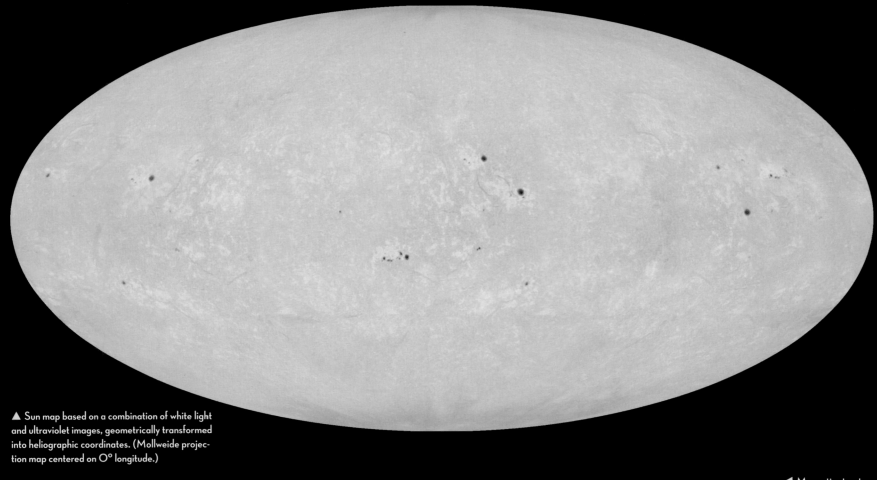

▲ Sun map based on a combination of white light and ultraviolet images, geometrically transformed into heliographic coordinates. (Mollweide projection map centered on 0° longitude.)

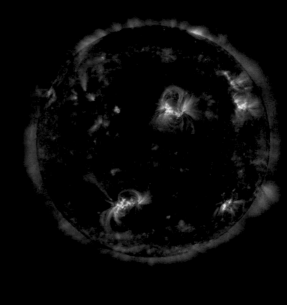

◀ Magnetic structures in the Sun's corona are revealed in this ultraviolet image, sensitive to a temperature of about 1 million Kelvin.

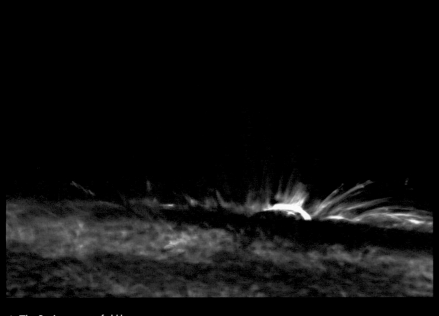

▲ The Sun's magnetic field becomes stronger around sunspots. This image shows the vertical structure of ionized gas following the local magnetic field lines at the edge of a sunspot on the Sun's limb.

▶ An immense coronal mass ejection (CME) erupts through the sun's coronosphere in this image from January 4, 2002.

◀ This view shows a prominence near the Sun's south pole. A combination of three ultraviolet wavelengths (at 30, 171 and 195 Angstroms) gives the view its color.

▲ This image combines an ultraviolet satellite view of the Sun's surface with a photograph of the corona taken from the Earth during an eclipse.

▲ Different wavelengths of ultraviolet light are used to explore structures within the sun's atmosphere.

▲ Waves and loops of ionized gas play around the sources of several solar flares on March 30, 2010.

What is it made of?

CECILIA PAYNE WROTE the most important astronomy PhD thesis of the 20th century, yet hardly anyone knows her name. In the 1920s she discovered that the Sun is 98 percent hydrogen and helium, gases essentially absent on Earth. So controversial was this that she wrote in her thesis that the abundance of the two gases is "improbably high and almost certainly not real." Years later, when the evidence supporting her was overwhelming, it was her supervisor, Henry Norris Russell, who got the credit.

Payne's result was controversial because everyone at the time believed the Sun was made of iron. In the 19th century, German scientists had found that atoms, when heated, glowed with light of characteristic colors, or wavelengths. Every element–oxygen, hydrogen, calcium, gold–had a unique fingerprint. The new science of spectroscopy detected in the Sun's spectrum the overwhelming fingerprint of one element: iron.

Atoms give out or absorb light when an electron jumps from one orbit around a "nucleus" to another. Payne's crucial insight was that, at the high temperature of the Sun, with atoms flying about and colliding violently, some atoms lose most, if not all, of their electrons. This is the case for hydrogen and helium, which have one and two electrons respectively. These elements therefore punch far below their weight in the light of the Sun. By contrast, iron, with 26 electrons, hardly ever loses all of them. Because of this it is prominent in sunlight. It was when Payne compensated for this effect using the newfangled "quantum theory" that she deduced that hydrogen and helium were superabundant.

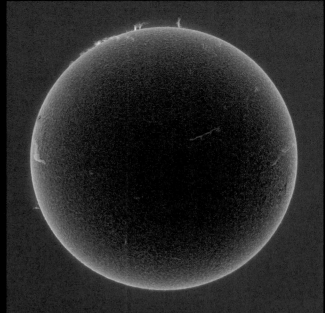

▶ Ionized hydrogen gas in the atmosphere of the Sun is highlighted by using a filter in the deep red end of the visible spectrum.

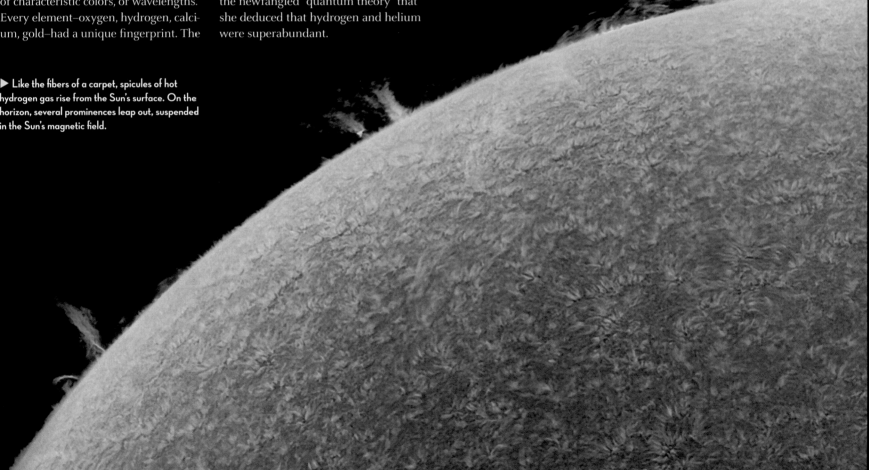

▶ Like the fibers of a carpet, spicules of hot hydrogen gas rise from the Sun's surface. On the horizon, several prominences leap out, suspended in the Sun's magnetic field.

Why is the Sun hot?

THE SUN IS HOT because it contains a lot of mass. Simple as that. All that mass weighs down on the solar core. If you have ever used a bicycle pump, you will know that squeezed air gets hot. The interior of the Sun is hot for the same reason.

So much matter is crushing down on the center of the Sun that the temperature there is about 15 million degrees Celsius. At such high temperatures, all matter–regardless of its identity–dissolves into an electrically charged gas, or "plasma." The key thing is that this is an anonymous, uniform state. So it makes little difference what the Sun is made of. It ends up in a very similar state.

The Sun is about a billion billion billion tons of mostly hydrogen gas. But if you put a billion billion billion tons of microwave ovens in one place, or a billion billion billion tons of bananas, you would end up with something as hot as the Sun and like the Sun.

OK, not exactly like the Sun. Although the central temperature of the Sun depends only on the amount of matter it contains, what it's made of plays a secondary role because free electrons hinder the escape of heat. The more electrons in an atom–and heavy atoms like iron have more than light ones like hydrogen–the more effectively heat is bottled up inside the Sun.

So much for why the Sun is hot, but why does it stay hot?

▲ There are only eight bananas here, so they don't have quite as much mass as the sun, but imagine how many would be needed to make a billion billion billion tons.

Why does it stay hot?

THE SUN IS CONTINUALLY losing heat into space, so it should be cooling. But it is not. Evidently, something is replacing the heat as fast as it is lost, maintaining the temperature dictated by the Sun's mass. But what?

The question cannot be answered without knowing how much heat the Sun is generating. In the early 19th century, this was measured independently by Claude Pouillet and John Herschel, the latter from an island surrounded by hippo–infested marshes that is now a suburb of Cape Town called Observatory.

In the steam-powered 19th century, it was natural to speculate that the Sun was powered by coal. How long could the Sun maintain its measured heat output if it were a lump of coal? The answer was barely 5,000 years. However, evidence from geology and biology was that the Earth was far older. Today's estimate is almost 5,000 million years. So, whatever the energy source of the Sun, pound for pound it must be a million times more concentrated than coal.

In the 20th century, such an energy source was discovered: nuclear energy. At the high temperature in the Sun, the cores, or nuclei, of the lightest element, hydrogen, collide and stick, gradually building nuclei of the next heaviest element, helium. It is a very inefficient nuclear reaction. On average it takes two hydrogen nuclei 10 billion years to collide and stick. This is why the Sun can burn for billions of years, long enough for the evolution of complex life.

The Sun's nuclear reactions unleash a damburst of energy, which emerges as sunlight at the surface. But what do we mean by "surface"?

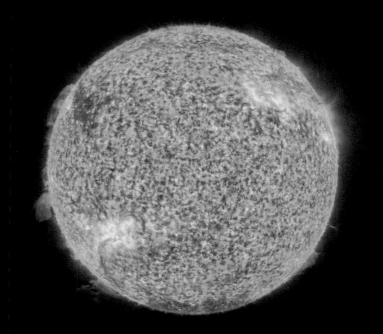

▲ One of the largest eruptive prominences of recent years rises from the Sun's limb. Such eruptions are triggered as the magnetic fields that hold the gas in the Sun's corona suddenly reconfigure.

Does the Sun have a surface?

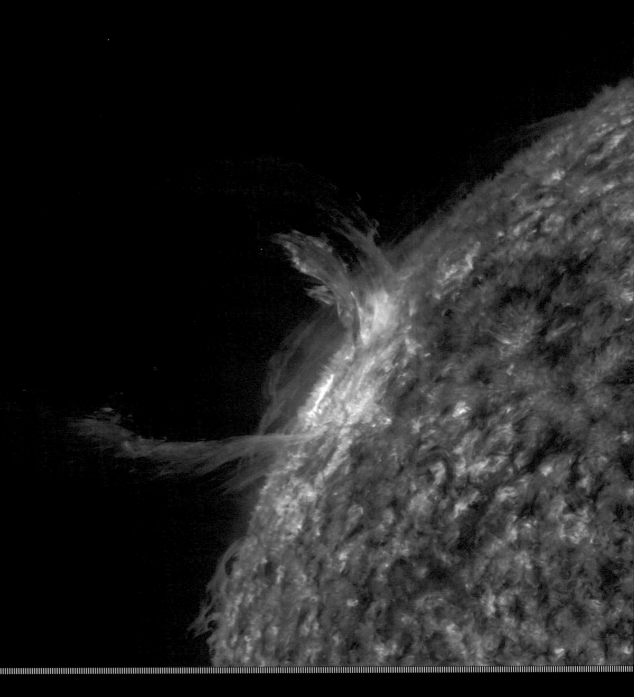

THE SUN IS about 300,000 times as massive as the Earth. Although the tremendous gravity of all this matter squeezes the deep interior so that it is far denser than any solid, the Sun is nevertheless a ball of gas. How, then, can it have a surface?

The answer is that it cannot–at least, not like the Earth's solid surface. Instead, it is defined by light.

Think of sunlight, generated by nuclear reactions at the core and working its way out. Free electrons get in its way like unhelpful pedestrians, so the light never travels more than a centimeter before being deflected. It zigzags its way out. So tortuous is this "random walk" that it takes about 30,000 years. Consequently, today's sunlight was made during the last ice age.

The Sun's "photosphere," or "surface," is where light working its way out of the Sun goes from walking to flying. Once it's outside this surface, sunlight is free to fly in a straight line and takes only about eight minutes to reach the Earth.

If light could travel in a straight line from the center the Sun, it would take two seconds, not 30,000 years, to get out. This is the time it takes "neutrinos," which are also created in the energy-generating nuclear reactions that make sunlight.

Hold up your thumb. About 100 million million neutrinos are slicing through the tip every second. Around eight minutes ago they were in the heart of the Sun. An amazing image was made by the Japanese Super-Kamiokande d the Sun, taken at night, looking down through the Earth, not with light but neutrinos.

But there is more to the Sun than neutrinos and sunlight. There is magnetism.

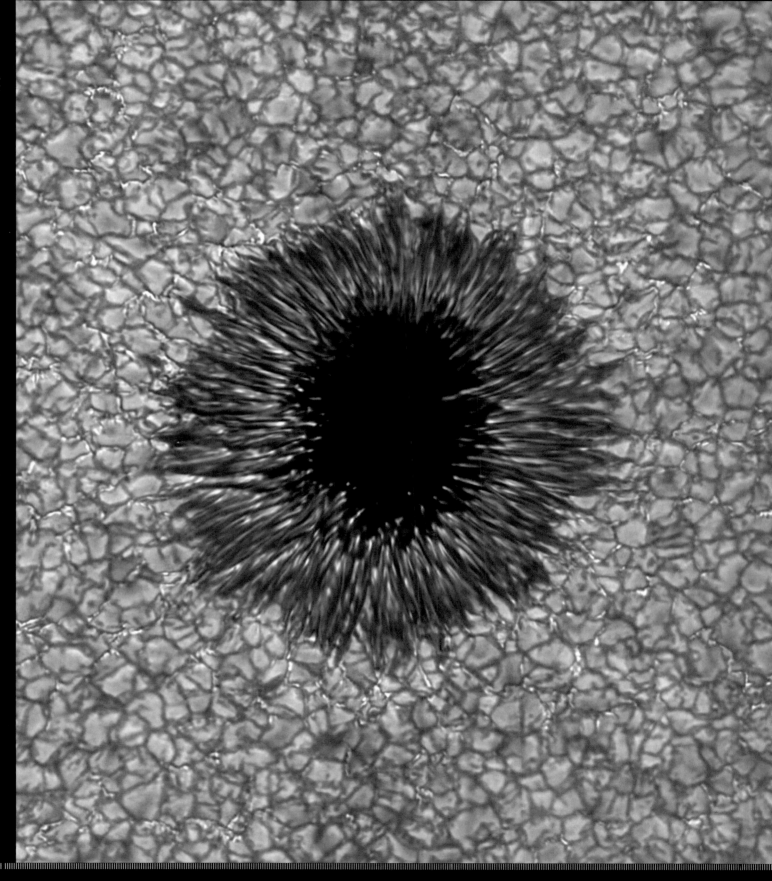

► The dark heart of a sunspot is shown in incredible detail in this image from a ground-based telescope. Surrounding the spot is a continuous pattern of convection cells.

Explosion on the Sun

IT IS SEPTEMBER 1859. Ships at sea are reporting tremendous blood-red "auroras"–shifting curtains of light in the night sky. Compasses are going wild. Telegraph operators are being electrocuted by their own equipment. One man knows what is causing all this, but no one will believe him.

On September 1st, Richard Carrington is observing the Sun from his observatory in Redhill, south of London, when he sees a bright explosion above a group of sunspots at the center of the Sun. Simultaneously, at Kew in London, the needle of a magnetometer goes off the scale. Learning of the coincidence, Carrington concludes a storm has erupted on the Sun–a storm that has reached out across space and engulfed the Earth.

This is scientific heresy. Carrington is ostracized by scientists of his day. Ever since Newton it has been the orthodoxy that just one force affects the Earth and planets: gravity. Sadly, Carrington will die without his discovery being fully accepted.

But what Carrington has confirmed is that the Sun is magnetic, and that its magnetism profoundly affects the Earth. The Earth does not hang in splendid isolation but is buffeted by cosmic events.

The solar "flare" of 1859 was the biggest ever recorded. If it occurred today, says Stuart Clark in *The Sun Kings*, electrical currents would be induced in power lines and electricity-generating stations sufficient to melt them. Satellites, computers, and communications networks would be destroyed. We would be returned to the steam age.

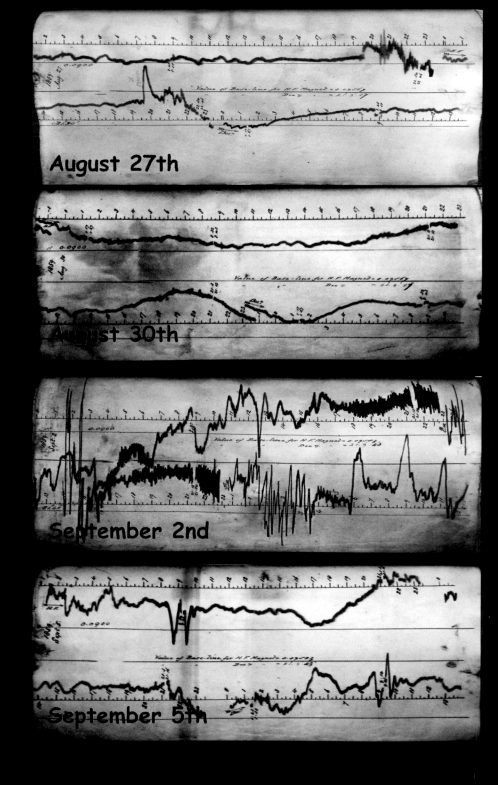

August 27th

August 30th

September 2nd

September 5th

▶ Carrington event. Magnometer readings from August 27, 1859 - September 7, 1859.

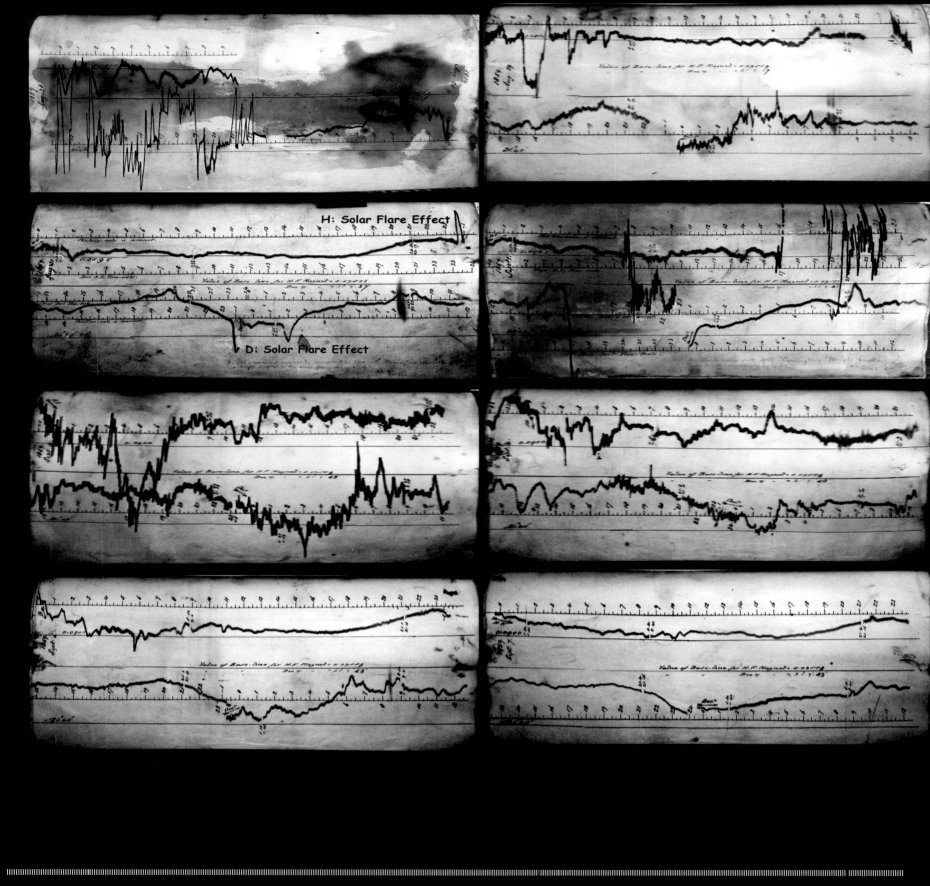

H: Solar Flare Effect

D: Solar Flare Effect

Magnetic Sun

THE SUN IS the strongest magnet in the Solar System. This manifests itself in many phenomena, most famously "sunspots." Dark splotches, often bigger than Earth, these are places where particularly intense magnetic fields break the surface.

The number of sunspots grows and declines every 11 years. This sunspot cycle is related to changes in the Sun's global magnetic field, which reverses every 11 years. The magnetic north pole becomes the magnetic south pole, and vice versa.

From 1645 to 1710, mysteriously, the Sun had very few sunspots. Known as the "Maunder Minimum," this period coincided with the middle and coldest part of the "Little Ice Age" in Europe and North America, a period of bitterly cold winters.

In addition to sunspots, the Sun exhibits flares when field lines become twisted like rubber bands and catapult matter into space. The biggest events, when huge quantities of matter are ejected, are known as "coronal mass ejections," a good example of which was the 1859 Carrington event.

The Sun's magnetic field is thought to be created by electrically charged currents of gas circulating in its interior. By rights, such a dynamo should run down, its currents slowing as they lose energy to their surroundings. However, a combination of solar rotation and hot matter "convecting"–welling up from below–appears to keep the magnetic dynamo going.

Even without flares, the Sun reaches out and touches the Earth–with the solar wind.

▲ The Sun's magnetic field deduced from observations of the Solar and Heliospheric Observatory (SOHO).

▶ At times of high solar activity, the Sun's surface is peppered with dark sunspots. Sunspots develop within brighter, hotter regions known as faculae. This image has been enhanced to show faculae as white areas.

The solar wind

THE SOLAR WIND is a million-mile-an-hour hurricane that blows out from the Sun and past the planets. It consists mostly of hydrogen nuclei and carries with it the Sun's magnetic field.

The wind's origin is not completely understood. Although the surface temperature of the Sun is less than 6,000 degrees Celsius, the Sun is surrounded by an atmosphere or *corona* at a temperature of millions of degrees. This is thought to be heated by shock waves from the Sun's boiling surface. Coronal gas particles are moving so fast, they can easily escape the Sun's gravity. Since the wind is the Sun's outer atmosphere and the wind extends to Earth and beyond, we are actually inside the Sun.

The solar wind takes about four days to cross space to reach us. Fortunately, the Earth has a magnetic field, much like a bar magnet, and this "magnetosphere" shields the planet so that the solar wind passes harmlessly around it like a stream around a boulder.

But particles of the solar wind can funnel down the magnetic field lines at the Earth's poles. There, they smash into air atoms, energizing their electrons, which, when they shed their surplus energy as light, create the colorful light displays of auroras.

Eventually, the solar wind slams into the interstellar gas, backing up on itself in a turbulent region known as the "termination shock." Beyond this is the calm of the "heliopause," where our sun's contribution blends in with the stellar winds of all the surrounding stars to form the interstellar medium. Mankind's most distant creation, Voyager 1, is expected to cross into the calm of true interstellar space, beyond the heliopause, some time in 2014.

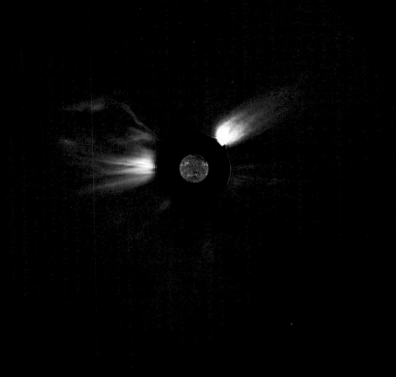

▲ It's not a DVD, but superhot coronal gas streaming away from the Sun, as observed by the Solar Terrestrial Relations Observatory (STEREO) satellite.

The death of the Sun

WHAT GETS HOTTER as it loses heat? The Sun.

Here is how it happens. In its core, the Sun is taking nature's basic Lego brick, the hydrogen nucleus, and assembling it into nuclei of helium, the by-product being sunlight. Helium, being heavier than hydrogen, falls to the center of the core. Its self-gravity squeezes it fiercely—and, when a gas is squeezed, it gets hotter.

So, paradoxically, the Sun gets hotter and brighter as it ages. It is now about 30 percent brighter than when it was born, posing a mystery: Why did the Earth not freeze into a giant snowball, never to recover?

In the future, as helium "ash" continues to rain down onto the center of the core, the Sun will get steadily hotter. In effect, it will become two stars in one: a tiny, white-hot core, deep inside a cool envelope inflated to monstrous size by heat flooding out from the core. It will be a "red giant"—a peach of a star.

So, will the Sun swallow the Earth? It depends. Red giants hold onto their outer regions only loosely—they shed matter into space. The Sun will certainly expand to the Earth's current orbit. However, the Earth, experiencing a weaker pull from a less massive Sun, may move out of reach.

The Sun's red-giant phase will be short–lived. In about 5 billion years' time, when it has spent all its hydrogen fuel, the Sun will settle down as a "white dwarf," a super-compact stellar ember about the size of the Earth. It will slowly fade, ending its life not with a bang but a whimper.

▲ When the sun swells to a red giant, it will develop a peach-like interior, with a super-dense core surrounded by a rarefied envelope.

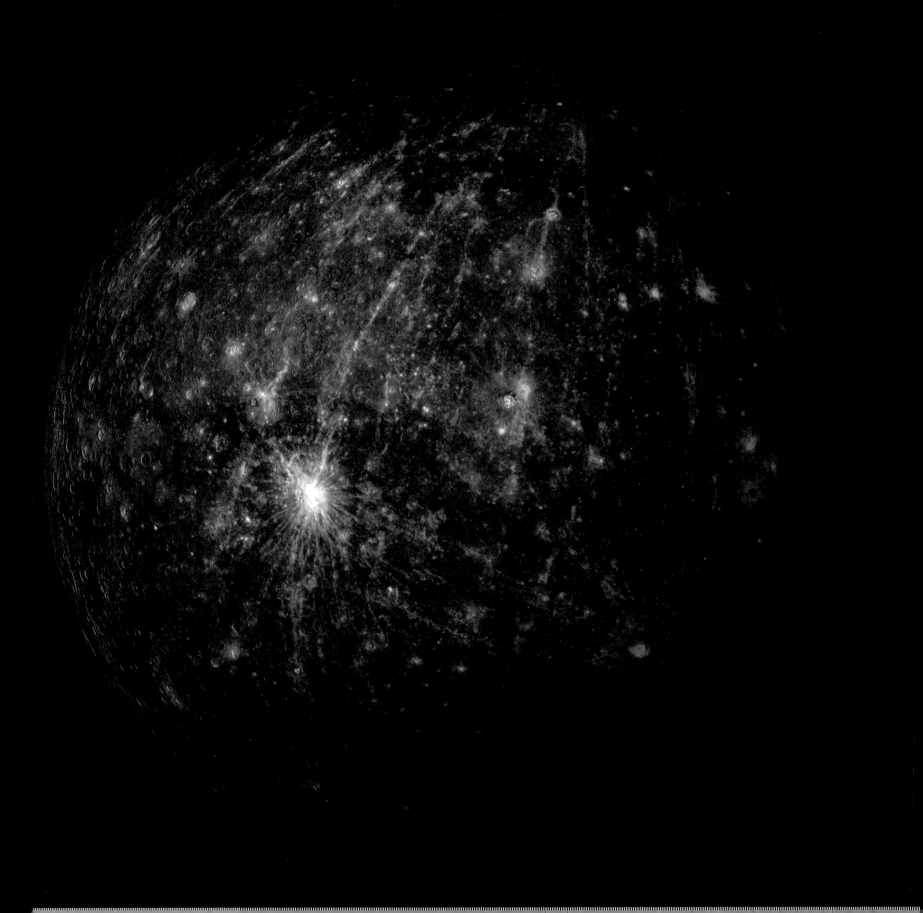

Mercury

MERCURY, THE INNERMOST planet, is one of the most boring objects in the Solar System. The equivalent of ten Suns in the skies of Earth blaze down on its scorched and crater-strewn surface. Lacking the protective shield afforded by an atmosphere or a strong magnetic field, Mercury is exposed to a deadly sleet of solar particles. Despite its alien harshness this tiny planet–smaller even than Jupiter's moon Ganymede–has more similarities with Earth than differences.

ORBITAL DATA

Distance from the Sun 46 to 70 million km / 0.31 to 0.47 AU
Orbital Period (Year) 87.97 Earth days
Length of Day 58.8 Earth days
Orbital Speed 59.0 to 38.9 km/s
Orbital Eccentricity 0.2056
Orbital Inclination 7°
Axial Tilt 0.5°

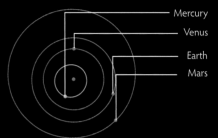

- Mercury
- Venus
- Earth
- Mars

PHYSICAL DATA

Diameter 4,874 km / 0.38 x Earth
Mass 330 billion billion metric tons / 0.06 x Earth
Volume 60,900 million km³ / 0.06 x Earth
Gravity 0.378 x Earth
Escape Velocity 4.251 km/s
Surface Temperature 100° to 700° K / −173° to 427° C
Mean Density 5.43 g/cm³

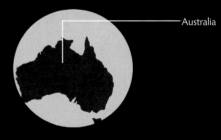

Australia

ATMOSPHERIC COMPOSITION

Hydrogen 99%
Helium 1%

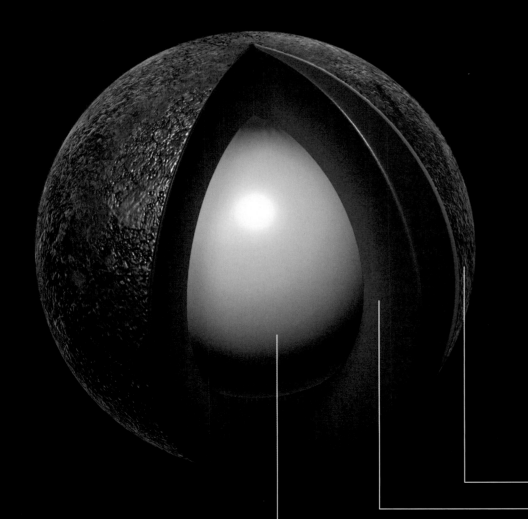

Rocky crust

Silicate mantle

Iron core

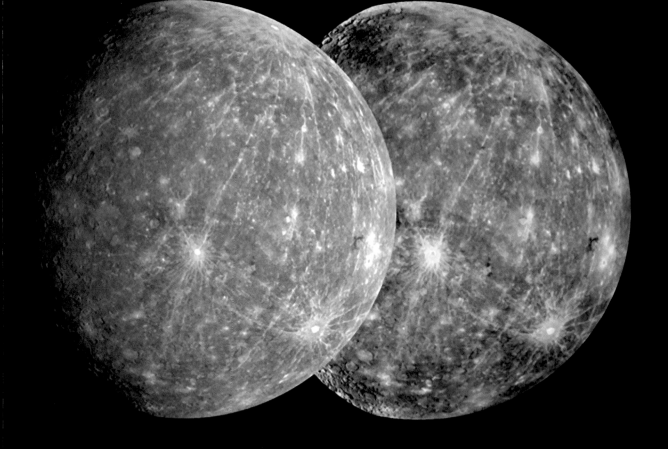

Two tribes

MERCURY IS a "terrestrial" planet. This needs some explanation. All the worlds orbiting close to the Sun–Mercury, Venus, Earth, and Mars–are tiny balls of rock, while all the worlds far away–Jupiter, Saturn, Uranus, and Neptune–are giant globes of gas. This difference reflects their origin 4.55 billion years ago in the debris disk swirling around the newborn Sun.

The disk was essentially made of hydrogen and helium gas pervaded by chunks of ice, silicate rock, and iron, of which meteorites are the remnants. Close to the Sun it was too hot for ice to survive, so the kilometer-sized "planetes-imals" that collided and stuck during the last stages of planet formation were rock and iron. Because the resulting planets were initially molten, the dense iron

sank to their centers, creating the iron cores of the terrestrial planets. Being so small, their gravity was too weak to gather about them a thick mantle of gas.

Contrast this with the situation far from the Sun, where it was cold enough for ice to survive. The planetesimals that stuck together were made of rock and iron–and abundant ice. They therefore grew many times bigger than terrestrial worlds. In fact, once they got to five to 10 times the Earth's mass, their gravity was strong enough to gather about them massive mantles of gas. The gas giants had to do this very quickly, before the newborn Sun "switched on" energy-generating nuclear reactions and blew away the gas in the disk. How this was possible remains a mystery.

▲ The surface of Mercury is a fairly uniform gray at visible wavelengths (right), but combinations of visible and near-infrared light can be used to highlight subtle color differences (left), helping geologists to map areas of different composition.

▼ Long shadows highlight Mercury's rough pock-marked surface in this oblique view. The large flat-floored crater Stravinsky, diameter 190 kilometers, crosses the right edge of the frame.

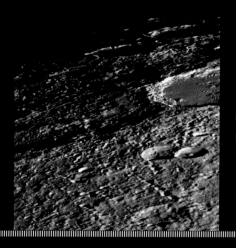

Fire and ice

ONCE UPON A TIME, it was thought that Mercury was "tid-ally locked"—that is, it kept one face always to the Sun, just as the Moon keeps one face to the Earth. This led to the idea that, paradoxically, one of the coldest places in the Solar System might be the permanently dark face of Mercury, the planet closest to the Sun.

Now we know that Mercury in fact turns on its axis once every 59 days, exactly two-thirds of the 88 days it takes to go around the Sun. Nevertheless, the planet's lack of an atmosphere to circulate warmth around the world means there are extremes of heat and cold. Deposits of ice exist near the north pole in the permanent shadows of deep craters. The source of the ice is believed to be comets that impacted the planet in the past.

Whether or not a planet can retain an atmosphere, per-haps of gases "burped" from its interior, depends on several factors. The closer to the Sun and the hotter a planet, the faster gas molecules fly about and the harder it is for a planet's gravity to grip onto them. Only a very massive body could do so. Mercury, alas, is far too small. Add to that its lack of a magnetic field to shield it from the ferocious solar wind and it is no surprise that it has no atmosphere to speak of.

◄ A crescent Mercury seen by the Messenger probe as it approached from the nightside of the planet in 2008.

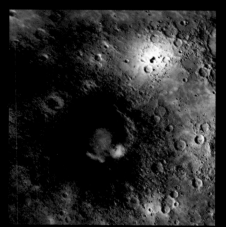

◄ Smooth plains cover large parts of Mercury's surface, suggesting widespread volcanic outpour-ing in its geological past. This false color view highlights in yellow the possible site of an explosive volcanic plains on the floor of the central crater.

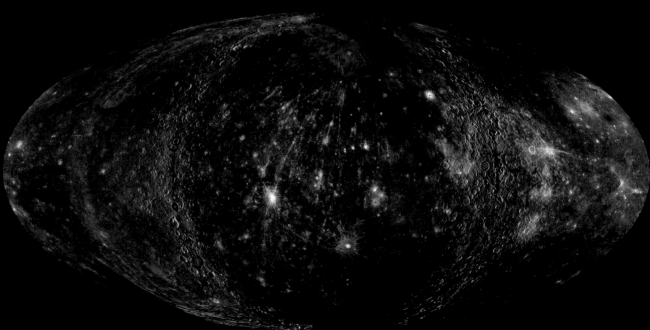

► Mercury map based on incomplete coverage from the first three flybys of NASA's MESSEN-GER probe (2008 and 2009) and Mariner 10's single flyby (1974). (Mollweide projection map centered on 0° longitude.)

How do planets move?

IT WAS THE POLISH astronomer Nicolaus Copernicus who realized that the planets go around the Sun rather than the Earth. Later, the German Johannes Kepler deduced, from a detailed analysis of observations, that those orbits were not circles, as many believed, but ellipses. The question was: why?

Isaac Newton's genius was to realize that each mass is attracted to every other mass. By comparing the effect of this "universal force" on a falling apple and on the Moon–which, Newton realized, was also falling in a circle–he deduced that this force of "gravity" weakened with an inverse-square-law. So, if two masses were moved twice as far apart, the force between them became four times weaker; three times as far apart, nine times weaker.

Crucially, Newton proved that the path of a planet gripped by a force obeying an inverse-square-law force is an ellipse. (Actually, this is not completely true. If a body were boosted ever faster, it would eventually escape, following a hyperbolic orbit. This is what happens when a space probe is launched into interplanetary space from the cargo bay of the Space Shuttle.)

Planets do not orbit in perfect ellipses because, in addition to being pulled by the Sun, they are tugged gently by the gravity of other planets. This causes their orbits to slowly change their orientation, or "precess," so they trace out a rosette. It is a tiny effect. The peculiar thing about Mercury, however, is that even if the pull of the other planets were to be magically removed, its orbit would still trace out a rosette. This anomaly baffled everyone–until it was explained by Einstein.

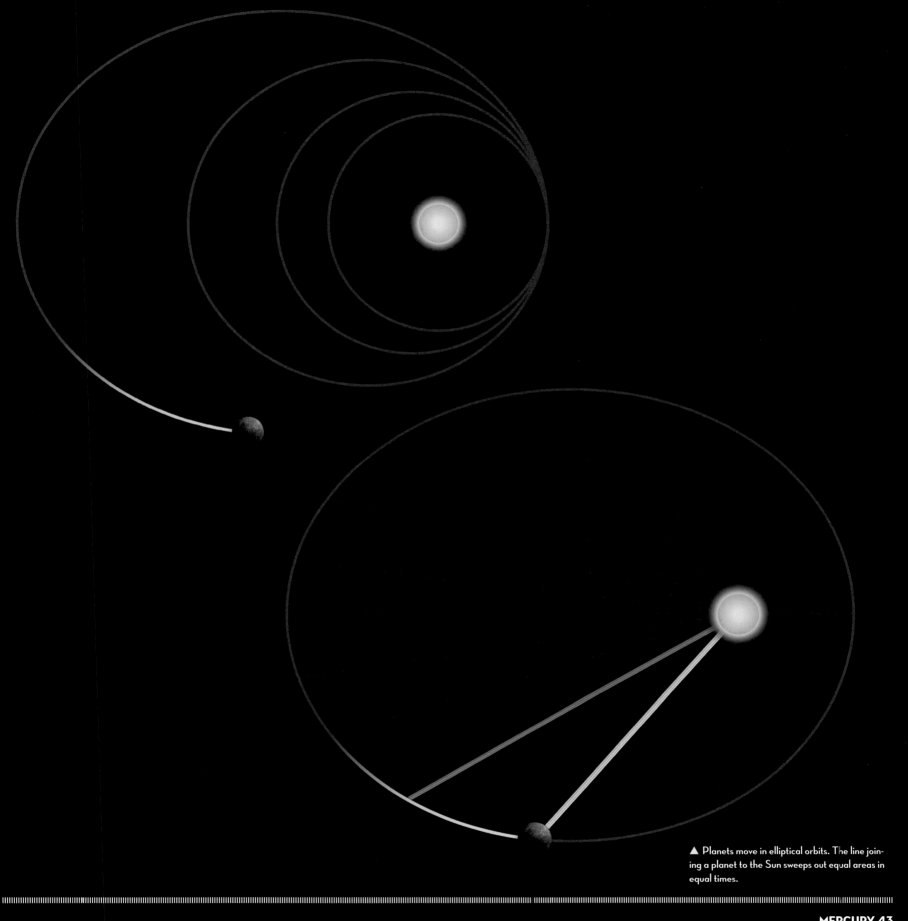

▲ Planets move in elliptical orbits. The line joining a planet to the Sun sweeps out equal areas in equal times.

Einstein vs. Newton

ACCORDING TO EINSTEIN, all forms of energy have mass—heat energy, light energy, the sound energy in your voice. Crucially, therefore, gravitational energy has mass too. Like all mass, it gravitates. Gravity creates more gravity.

The effect is tiny and appreciable only where gravity is strong—for instance, near the Sun. Mercury is near the Sun. Consequently it experiences slightly stronger gravity than Newton predicted. Since it is only under the influence of an inverse-square-law force that a planet traces out an ellipse, Mercury should be moving in a subtly different manner.

Einstein predicted that the orbit of Mercury should precess, tracing out a rosette that repeats about once every 3 million years. This had been observed and had baffled everyone. Einstein was triumphant. Newton, whose ideas had reigned for centuries, was wrong.

Mercury confirmed Einstein's theory of gravity, the general theory of relativity, published in 1915. Einstein's theory was bolstered by the bending of starlight near the Sun during the total eclipse of 1919.

In Einstein's theory, gravity is no longer seen as a "force" but rather as warpage of unseen four-dimensional space-time. Put simply, the theory says: "Matter tells space how to curve, and curved space tells matter how to move." It incorporates the idea of a maximum cosmic speed—that of light. Since gravity, like light, takes about eight minutes to travel from the Sun, if the Sun were to vanish it would be eight minutes before the Earth noticed.

Einstein's theory also predicts the existence of black holes; a universe that began in a titanic explosion, the Big Bang; and possibly even time machines.

▲ The Messenger space probe entered Mercury's orbit in March 2011. The spacecraft's instruments are protected from the Sun's intense glare by a sunshade of heat–resistant ceramic fabric.

▼ Mercury's orbit continually changes its orientation in space, gradually tracing out the pattern of a rosette.

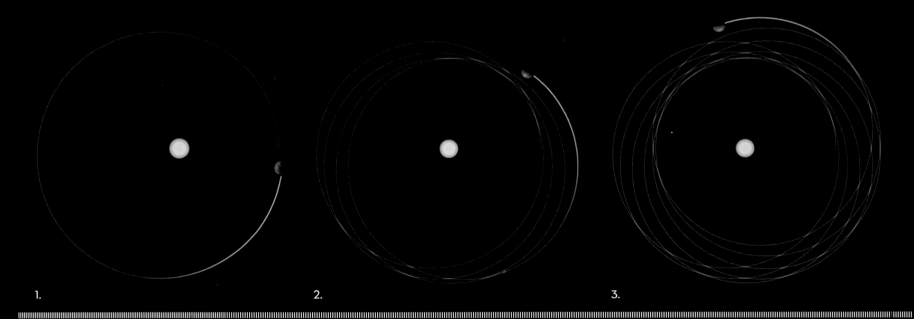

1.   2.   3.

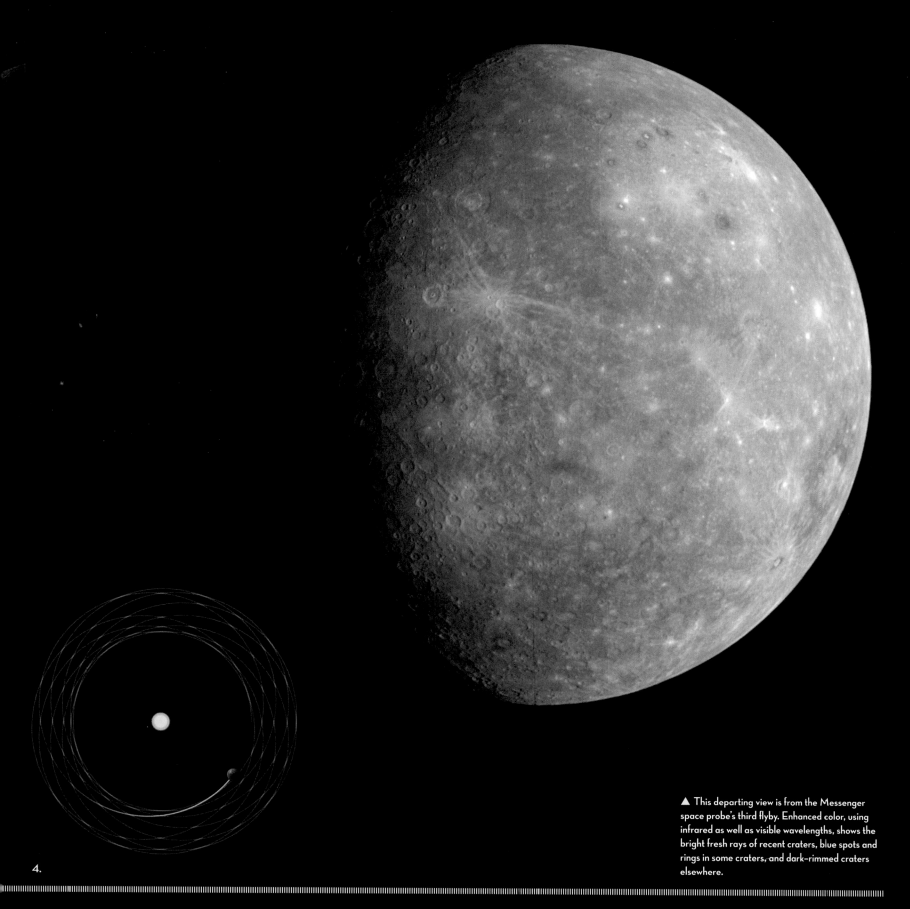

▲ This departing view is from the Messenger space probe's third flyby. Enhanced color, using infrared as well as visible wavelengths, shows the bright fresh rays of recent craters, blue spots and rings in some craters, and dark-rimmed craters elsewhere.

4.

Venus

VENUS IS HELL. There are no two ways about it. Beneath impenetrable sulfuric acid clouds lies a surface hot enough to melt lead. All space probes to have reached that surface have either instantly or after a short time been crushed like flies by the tremendous weight of the Venusian atmosphere, 100 times thicker than the Earth's. Yet, paradoxically, Venus is our twin. Its composition is similar to that of our planet and its mass is only marginally smaller. This striking similarity spurred science-fiction writers to imagine Venus as a steamy, jungle-infested world. So why were they so wrong? The simple answer is that Venus is closer to the Sun and therefore hotter than the Earth. This led to it losing its water to space and an escalating temperature driven by a catastrophic runaway greenhouse effect.

ORBITAL DATA
Distance from Sun 107 to 109 million km / 0.72 to 0.73 AU
Orbital Period (Year) 224.7 Earth days
Length of Day 243.02 Earth days
Orbital Speed 35.3 to 34.8 km/s
Orbital Eccentricity 0.0067
Orbital Incination 3.39°
Axial Tilt 177.3°

Mercury
Venus
Earth
Mars

PHYSICAL DATA
Diameter 12,104 km / 0.94 x Earth
Mass 4,870 billion billion metric tons / 0.82 x Earth
Volume 928,000 million km^3 / 0.82 x Earth
Gravity 0.905 x Earth
Escape Velocity 10.361 km/s
Surface Temperature 737°K / 464°C
Mean Density 5.25 g/cm^3

Earth

ATMOSPHERIC COMPOSITION
Carbon Dioxide 96.4%
Nitrogen 3.4%
Sulfur Dioxide .015%
Argon .007%
Water Vapor .002%

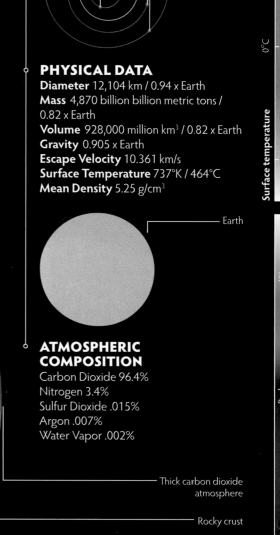

Thick carbon dioxide atmosphere

Rocky crust

Silicate mantle

Liquid metal outer core

Solid metal inner core

Surface temperature

800 K

400°C

600 K

200°C

400 K

100°C

0°C

200 K

0 K

Mean density

Iron

7 g/cm³

6 g/cm³

5 g/cm³

4 g/cm³

3 g/cm³

2 g/cm³

Rock

1 g/cm³

Water

0

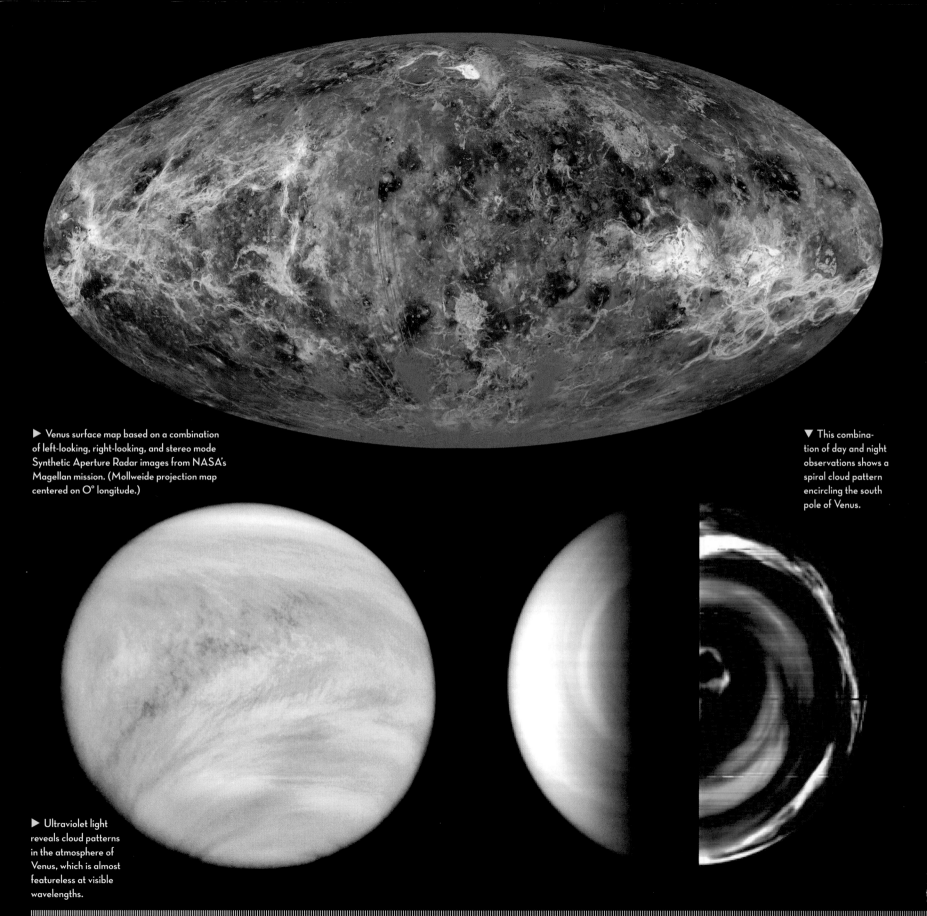

▶ Venus surface map based on a combination of left-looking, right-looking, and stereo mode Synthetic Aperture Radar images from NASA's Magellan mission. (Mollweide projection map centered on 0° longitude.)

▼ This combination of day and night observations shows a spiral cloud pattern encircling the south pole of Venus.

▶ Ultraviolet light reveals cloud patterns in the atmosphere of Venus, which is almost featureless at visible wavelengths.

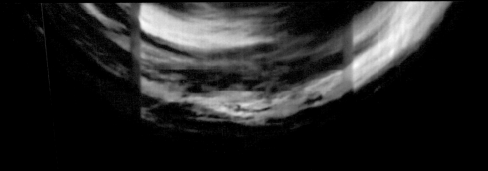

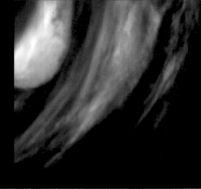

► Ultraviolet imagery shows stripes in the atmosphere of Venus, possibly due to to suspended dust and aerosols.

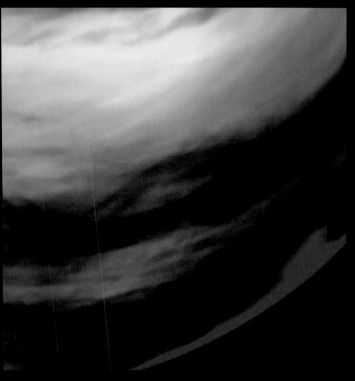

▲ A mosaic of three images taken of Venus's night side shows the thermal radiation emitted from clouds at an altitude of about 15–20 kilometers, as well as heat emitted from the planet's surface in the brightest areas.

▲ The "airglow" from oxygen atoms in the night-time atmosphere of Venus appears blue in this image. The flourescent glow is light emitted when oxygen atoms migrating from the day-to-night-side recombine into molecular oxygen.

▼ Color image from the surface of Venus taken by Russia's Venera 13 probe, which survived for 2 hours and 7 minutes after landing on March 1, 1982. The flat slabs of rock are thought to be similar to terrestrial basalt. The bright object on the ground is the camera's lens cover.

▲ Danilova is a large impact crater with a bright halo of ejecta and prominent central peak.

▼ A crescent Venus, close to quarter phase, viewed by the Pioneer Venus Orbiter in 1978.

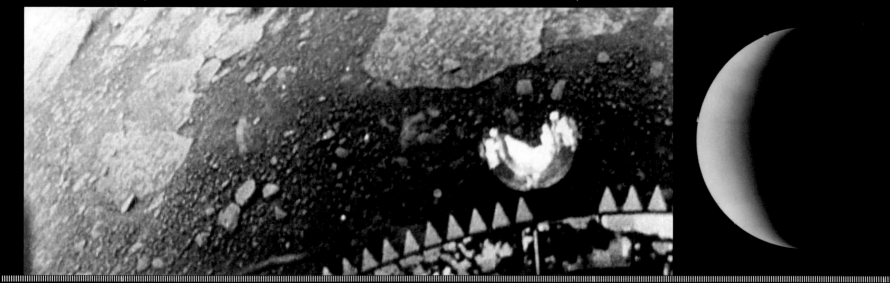

Hothouse world

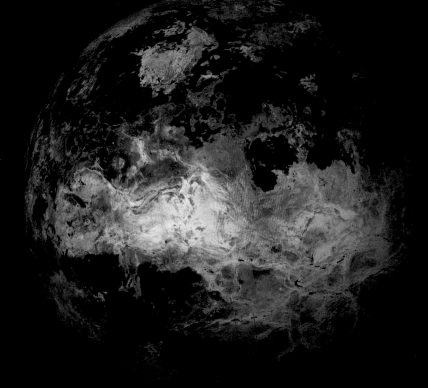

OCCASIONALLY, PEOPLE question why so much money is spent on planetary exploration. Venus provides one answer. The "hell" planet is a sober warning to us all. If we continue to pump carbon dioxide from the burning of fossil fuels into our atmosphere, Venus is our future.

Venus is believed to have started out like the Earth. But being nearer the Sun and hotter, water was driven from its oceans into its atmosphere. Water vapor is a greenhouse gas that traps heat. It boosted Venus's temperature, which evaporated more water vapor, which boosted the temperature yet more, and so on. Eventually, this runaway greenhouse effect led to the oceans boiling away entirely.

This was not the end of the climatic catastrophe. At the top of the atmosphere, ultraviolet light from the Sun split apart water molecules into their constituent hydrogen and oxygen atoms, gases that floated away into space. Without water, carbon dioxide, pumped out by Venus's volcanoes, could not be rained out. A greenhouse gas like water vapor, it built up, creating an atmosphere a hundred times denser than our own.

On Earth, if we continue to inject more carbon dioxide into the atmosphere, the temperature will go up and the oceans will begin to vanish, as on Venus. Earth has exactly the same amount of carbon dioxide as Venus, but locked up in chalk cliffs. If rising temperatures bubble all of that out into the air, the Earth will complete its transformation into Venus.

Besides the greenhouse effect, Venus's lack of water has other serious implications for the planet.

▼ Simulated perspective view of the Venusian surface around the volcano Sapas Mons. Bright lava flows extend for hundreds of kilometers across the fractured plains in the foreground.

▲ Once upon a time water up to 2 kilometers deep may have existed on Venus.

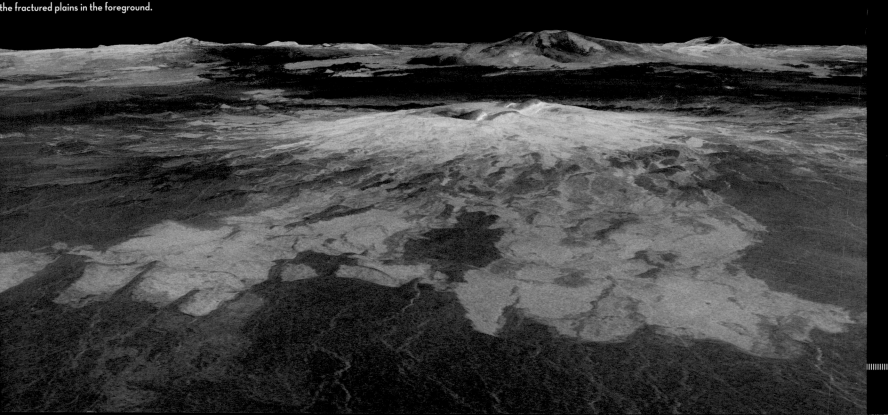

Cloud world

IN RAY BRADBURY'S 1954 story "All Summer in a Day," schoolchildren on Venus lock a classmate in a closet on the only day in seven years when the clouds are set to clear and the Sun to come out. In reality, it is not like that on Venus—it is worse. Venus is permanently shrouded in impenetrable sulfuric acid clouds. Frustratingly, this means we cannot see the surface of the nearest planet to Earth—except with radar.

Radio waves, unlike visible light, go through the clouds unhindered. If the echoes are picked up, it is possible to use them, with some computer wizardry, to build up a radar image of the hidden surface. This was triumphantly achieved by the "synthetic aperture radar" of NASA's Magellan spacecraft, which orbited the planet from 1990 to 1994.

Magellan revealed that Venus is not only a cloud world but also a volcano world, with an astonishing 50,000 volcanoes peppering the planet. It is the sulfur dioxide they spew out that makes the sulfuric acid clouds.

On Earth, the crust is thick and volcanoes break the surface only here and there—for instance, at tectonic plate boundaries or on the seafloor, which is thin. Venus appears to have no moving plates, only a globally thin crust that lava can puncture almost anywhere.

Why does Venus not have plate tectonics? It is yet another consequence of the planet losing its water. Water is the lubricant that oils the conveyor belt of plate movement.

▲ NASA's Magellan satellite carried a powerful radar to penetrate Venus' thick cloud cover. This image, about 630 kilometers across, shows lava flows that have crossed a ridge and pooled between the ridge and a mountain range in the lower part of the image.

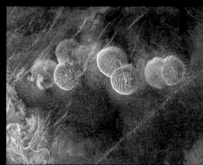

▶ A cluster of "pancake domes" on the eastern edge of alpha Regio. These dome-like hills, about 25 kilometers across and 750 meters high are peculiar to the planet Venus. They are likely formed by eruptions of very thick silica-rich lava.

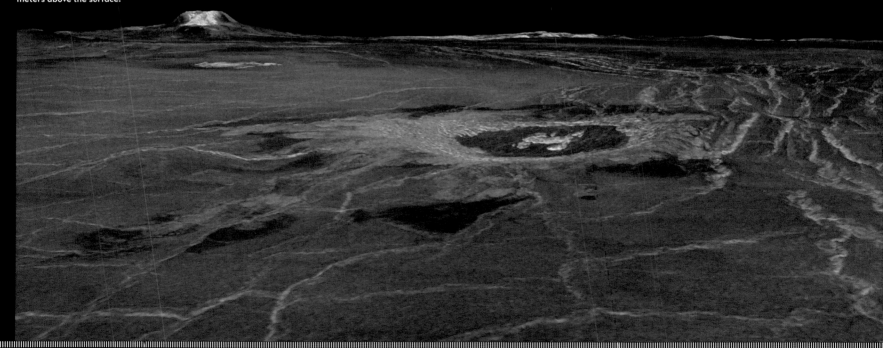

▼ Simulated view of the plains of western Eistla Regio, Venus. The impact crater Cunitz in the foreground is about 48 kilometers in diameter. On the horizon, Gula Mons rises to 3,000 meters above the surface.

Transit of Venus

IN 1769, Captain James Cook, one of the world's greatest explorers, arrived in Tahiti, which had been discovered by Europeans only the year before. The purpose of his voyage was to observe from the island the rare transit of Venus.

Roughly every 120 years it is possible to see the tiny black disk of Venus cross, or transit, the face of the Sun. In fact, this phenomenon occurs twice, eight years apart.

In Cook's day, such transits were crucial for astronomy. People knew the relative distances of the planets from the Sun. As Johannes Kepler had discovered, the orbital periods, which can easily be measured, are related in a simple way to the distances. For instance, Jupiter is five times as far away as the Earth. But although astronomers knew the relative distances, they did not know the absolute distances. This crucial information could be deduced from timing the passage of Venus across the Sun and applying some simple geometry.

In fact, Cook's measurements on June 3, 1769, were not very accurate. It was hard, for example, to determine the exact edge of a bright and fuzzy Sun. Nevertheless, it was a valiant effort to calculate the distance of the Earth and Venus from the Sun.

Today, the most accurate distance measurements come from radar—that is, using a radio dish to bounce a signal off a planet and timing how long it takes for the echo to return to Earth. Knowing the speed of radio waves, which is the same as that of visible light, it is possible to determine the distance.

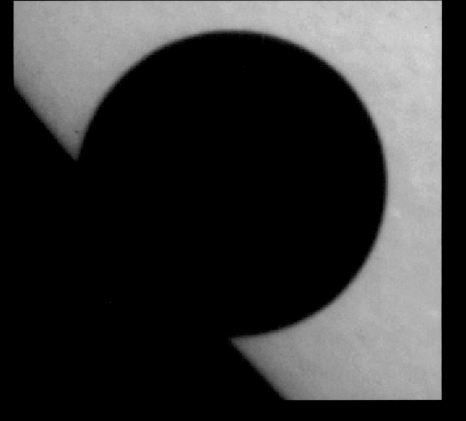

▲ This close-up image of the 2004 transit of Venus shows sunlight scattered within the thick atmosphere as a faint ring of light around the planet's dark disc.

◀ Venus passed in front of the Sun's disc when viewed from Earth on June 8, 2004. Transits of Venus are rare events since the orbit of Venus is tilted relative to the Earth's.

How to spot a planet?

VENUS IS BY far the easiest planet to spot in the night sky. A superbright white light, outshone only by the Sun and Moon, it is visible just before the Sun rises or after it sets. Hence it is commonly known as the morning or evening star.

Other planets are also easy to spot because they are generally brighter than stars. Mars is red, Jupiter white, and Saturn slightly yellow in color. Mercury, which appears orange, is difficult to spot because, being the closest planet to the Sun, it is near the horizon before sunrise or after sunset.

All the planets orbit pretty much in a single plane known as the "ecliptic." In practice this means they move in a narrow band around the night sky. People have joined the dots of the stars along this zodiac to create 12 signs, such as Cancer and Aries.

The word *planet* actually means "wanderer," a recognition by ancient sky watchers that there is something peculiar about them–they wander against the background of fixed stars, drifting along the zodiac. Actually, if you were to monitor a planet like Mars, night after night for months, you would see its path occasionally loop back on itself. This is because the Earth, which orbits the Sun faster than Mars, often catches up with the sluggish Red Planet, so from our point of view it drops behind us.

Planets are also unique in not twinkling like stars. Whereas turbulence in the atmosphere can cause a star to jump about or jitter, it is sufficient only to jitter the boundary of a much bigger planetary disk. This is not enough to cause its brightness to fluctuate significantly, or twinkle.

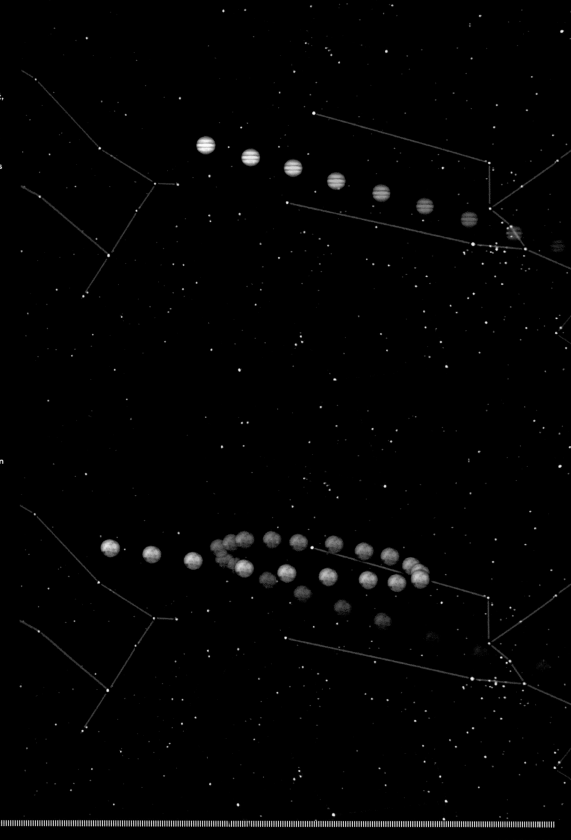

▶ From night to night, the planets move across the backdrop of stars. Here Jupiter is moving from the constellation of Taurus into Gemini.

▶ Planets farther out from the Sun than the Earth, such as Mars, occasionally appear to loop back on themselves as Earth "overtakes" them on its orbit.

Earth

IT IS BIG. It is round. And we all depend on it. When considering a world as familiar as Earth, it is hard to say anything new. But our planet is deeply mysterious. It is the only world with surface water. It is the only world with plate tectonics, an ozone layer–and life. Why is the Earth so special? It must be related to its distance from the Sun–smack bang in the Goldilocks zone, not too hot, not too cold–to its mass and composition, its giant moon, which stabilizes climate. Whereas the most complex thing on other planets is weather, on Earth complexity has increased remorselessly–from bacteria to multi-cellular life, human society, civilization, and technology. If you know why here and nowhere else, there is a Nobel Prize waiting for you. In the meantime, consider the ways the Earth is unique.

ORBITAL DATA

Distance from Sun
147 to 152 million km / 0.98 to 1.02 AU
Orbital Period (Year) 365.26 Earth days
Length of Day 23.935 Earth hours
Orbital Speed 30.3 to 29.3 km/s
Orbital Eccentricity 0.0167
Orbital Incination 0°
Axial Tilt 23.44°

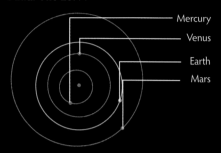

- Mercury
- Venus
- Earth
- Mars

PHYSICAL DATA

Diameter 12,756 km
Mass 5,970 billion billion metric tons
Volume 1,080,000 million km^3
Gravity 1 x Earth
Escape Velocity 11.18 km/s
Surface Temperature 204° to 331°K / −69° to 58°C
Mean Density 5.515 g/cm^3

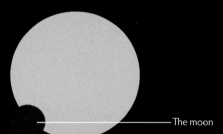

The moon

ATMOSPHERIC COMPOSITION

Nitrogen 78.084%
Oxygen 20.946%
Argon 0.9340%
Water vapor 0.1000%
Carbon dioxide 0.039%
Neon 0.001818%
Helium 0.000524%
Methane 0.000179%
Krypton 0.000114%
Hydrogen 0.000055%
Nitrous oxide 0.00003%
Carbon monoxide 0.00001%

Nitrogen/oxygen atmosphere

Seas and oceans

Rocky crust

Silicate upper mantle

Solid iron–nickel inner core

Liquid iron–nickel outer core

Silicate lower mantle

Surface temperature

800 K | 400°C
600 K
400 K | 200°C
200 K | 0°C
0 K

Mean density

Iron | Rock | Water
7g/cm³ | 6g/cm³ | 5g/cm³ | 4g/cm³ | 3g/cm³ | 2g/cm³ | 1g/cm³ | 0

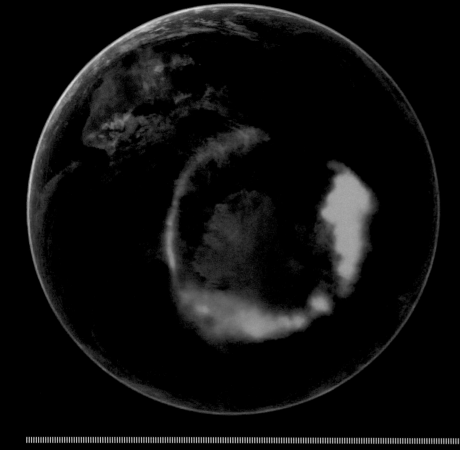

▲ Cloud-free Earth map based on a composite of thousands of images from the Advanced Very High Resolution Radiometer (AVHRR) on the NOAA TIROS polar-orbiting weather satellites. (Mollweide projection map centered on O° longitude.)

◄ Earth's strong magnetic field produces spectacular light shows—aurorae—around the North and South magnetic poles.

▶ The first fully illuminated picture of the whole Earth from space, taken by the crew of Apollo 17 on their way to the moon.

▶ The Southern Lights shimmer over the Indian Ocean, as charged particles from the Sun slam into the atmosphere, funneled by the Earth's magnetic field.

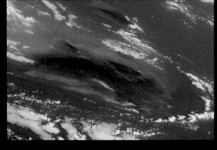

▲ Mauna Loa on the Big Island of Hawaii is the largest volcano on Earth, rising 9,700 meters from the floor of the Pacific Ocean. It is just one in a chain of volcanic peaks spread across the Pacific crustal plate as it moves over a hotspot in Earth's mantle. To see how large volcanoes can grow in the absence of plate tectonic motion, check out Olympus Mons on Mars.

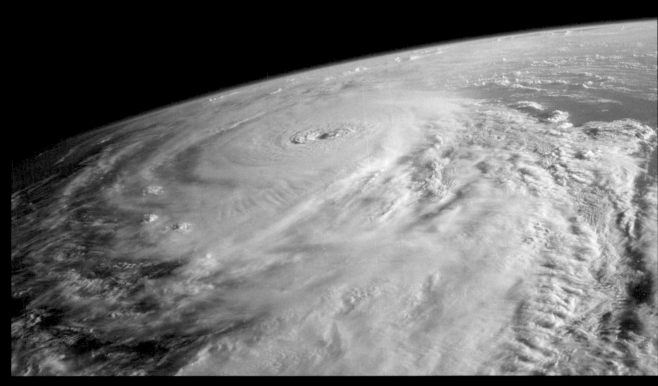

▶ Hurricane Lili over the Gulf of Mexico in 2006. Intense storms arise every summer over Earth's oceans, in the tropical latitudes just north and south of the Equator. Hurricane Lili was the deadliest storm of the 2002 Atlantic hurricane season, killing 15 people.

▼ Infrared wavelengths of light are used by geologists to gauge the chemical composition of rocks in remotely-sensed images. Limestone, sandstone, claystone and gypsum appear in different hues in this false-color satellite image of the Anti-Atlas Mountains of North Africa.

▼ Tides and currents at the edge of the Great Bahama Bank have sculpted sandbanks and seaweed beds into a work of abstract art. The bank is a vast undersea plateaus, mostly less than 10 meters deep, but sharply dropping away to 4,000 meters to the north.

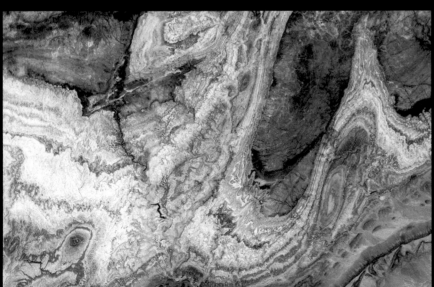

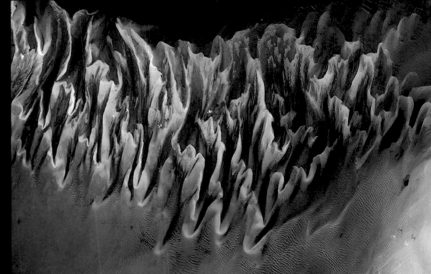

Water world

THE EARTH IS unique in being at a distance from the Sun where it is possible for liquid water to exist on its surface. More than 70 percent of the planet is covered in water, which in places is up to 11 kilometers deep.

Water is critical for life, providing the medium in which life's chemicals can come together and interact. The Earth's water may have originally been emitted in a "Big Burp," when heat released by radioactive rocks turned its interior molten, driving out liquids. However, there is evidence that some, possibly most, of the water was carried to Earth by comets that struck the planet in its youth.

The Earth came very close to being covered completely in water. If that had happened, human-level intelligence may never have arisen. In the oceans, after all, there has been no evolutionary pressure for whales and dolphins to evolve opposable thumbs with which to make tools and manipulate their environment.

Water's existence on Earth appears to be precarious. Repeatedly, the planet has been plunged into ice ages, triggered in part by variations in the Earth's orbit and axial tilt, known as "Milankovitch cycles." Since a frozen planet reflects solar heat back into space like a mirror, it is hard to understand how the Earth warmed again, particularly after a "Snowball Earth" phase, about 650 million years ago, when the planet was totally ice–covered. The warming of greenhouse gases, pumped out by volcanoes, is suspected to be crucial.

Besides making life possible, water, as rain, washes out sulfur dioxide from volcanoes, preventing the buildup of sulfuric acid clouds (see Venus). Water even lubricates the Earth's plates, oiling the conveyor belt of plate tectonics.

▲ A sight-seeing boat is just visible through the mist thrown up by Niagara Falls as the river's water plunges almost 50 meters along a 790-meter front.

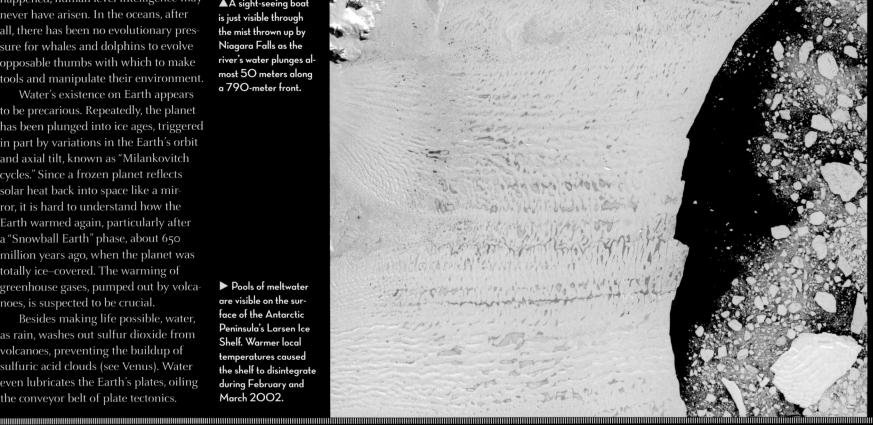

▶ Pools of meltwater are visible on the surface of the Antarctic Peninsula's Larsen Ice Shelf. Warmer local temperatures caused the shelf to disintegrate during February and March 2002.

▲ Earth's southern continent, Antarctica, is almost completely covered by a thick ice cap. The ice flows down to the continent's edge, where it breaks off to form icebergs, as seen here, where the Matusevich Glacier pushes through a channel between coastal mountain ranges.

▼ The mighty River Lena drains water from 2.5 million square kilometers of eastern Siberia, but splits into hundreds of tiny streams as it empties into the Arctic Ocean

◀ Alternating layers of dust and ice, accumulated over 10,000 years since the last ice age, are seen on the surface of the Barnes Ice Cap on Baffin Island, northern Canada. Similar layered polar deposits are seen around the north pole of Mars.

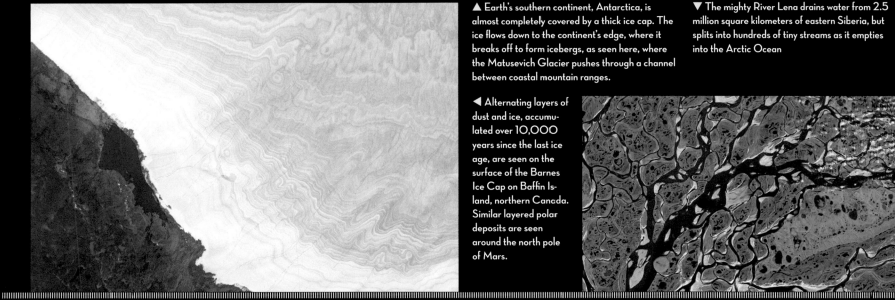

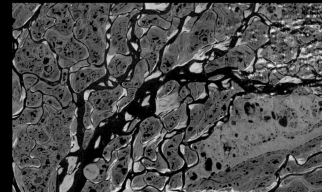

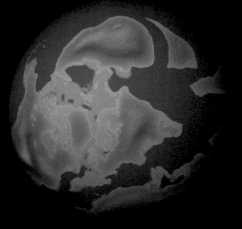

400 million years ago:
Earth's continents are moving south as ancient oceans close.

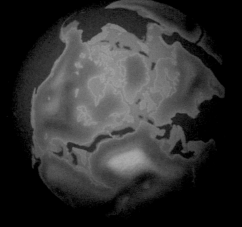

320 million years ago:
Today's South America, Africa, India, and Antarctica are grouped around the South Pole, forming the continent of Gondwana.

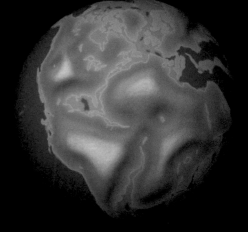

240 million years ago:
All the major landmasses are linked as the super-continent of Pangea stretches from North to South Poles.

Wegener's jigsaw

▶ A mushroom cloud of ash and steam punches a hole in the cloud deck as the volcano Sarychev erupts on Matua, one of the Kuril Islands extending south from Russia's Kamchatka Peninsula.

IN 1930 ALFRED Wegener died on a field trip in frozen Greenland. Consequently, he never got to see the triumph of his hugely controversial idea of "continental drift." Like Francis Bacon in 1620, Wegener noticed that the coastlines of Africa and South America would fit together like jigsaw pieces. Could the continents have been joined, then drifted apart? Wegener's idea has been incorporated into the modern science of plate tectonics.

The Earth's solid outer layer, or "lithosphere," floats on molten magma. There are two types of lithosphere: "oceanic crust," which is thin and dense, and "continental crust," which is thick and light and so floats higher. Crucially, the lithosphere is broken up into irregular fragments called "plates."

Where two continental plates collide, the crust buckles up into mountains like the Himalayas. Where a light continental and a dense oceanic plate collide, the oceanic plate dives beneath, buckling rock on the plate above to make mountains like the Andes, with the frictional heat creating volcanoes. Where the plates pull apart, at "mid-ocean ridges," lava wells up to fill the gap, creating new crust. Remarkably, a new ocean is today being born where three plates are pulling apart at Afar in Ethiopia.

The driving force of plate tectonics is hot magma rising and cool magma sinking in the Earth. Heat comes from the radioactive decay of uranium, thorium, and potassium in the Earth's rocks. In fact, it was radioactive heating that initially melted the Earth's interior, allowing dense iron to sink to the core and lighter rock to float to the surface and form the lithosphere.

Besides rock and water, the Earth of course has air.

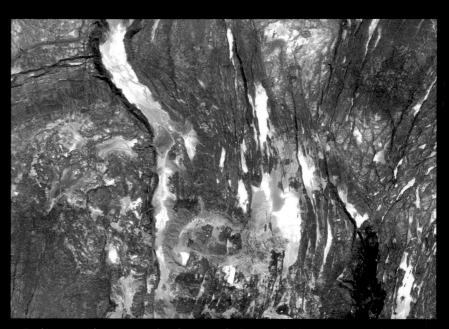

▲ Earth's crust is splitting apart in the Afar region of Ethiopia, at the three-way junction of the Red Sea, the Gulf of Aden and the Great Rift Valley of East Africa. The newest parts of Earth's crust show as dark lava flows, destined one day to be the floor of a new ocean, while bright sand has accumulated in older rifts.

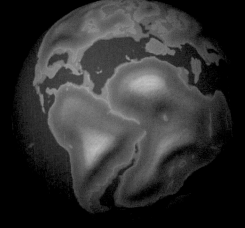

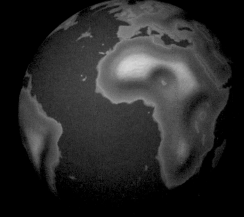

160 million years ago:
The Central Atlantic Ocean opens, pushing North America away from Europe, Africa and South America, heralding the breakup of Pangea.

80 million years ago:
The South Atlantic and Southern Oceans are opening up, splitting Gondwana into today's continents of South America, Africa, Antarctica and Australia.

Present day
Today the Atlantic Ocean separates the Americas from Europe and Africa.

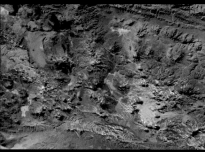

▲ Volcanic cones in the Andes Mountains are the result of the melting of the Pacific oceanic plate as it is pushed down into Earth's mantle under the South American continental plate.

► Earth has one of the most active surfaces in the Solar System. Its rocks, laid down in layers over millions of years, have been tilted, folded, and uplifted by compressive forces, then stripped back and sculpted by water and wind erosion, as seen here in the Anti-Atlas Mountains of North Africa.

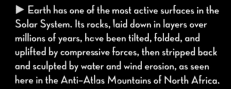

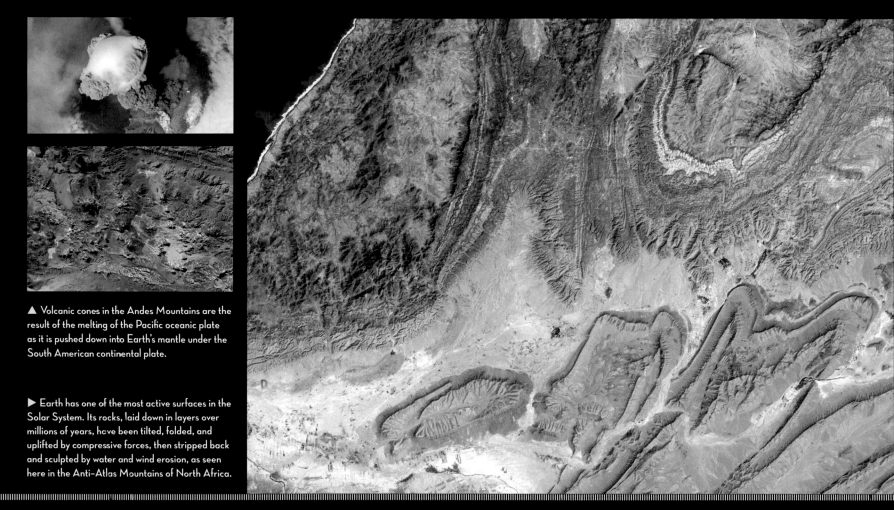

Earth's aura

FROM SPACE the Earth's atmosphere looks impossibly thin, barely as thick, if the planet were shrunk to the size of an apple, as the apple's skin. Nevertheless, the atmosphere is a cauldron of activity, energized by solar heat.

Variations in the sunlight falling on each hemisphere as the Earth orbits the Sun create seasons. This is because spinning things like the Earth tend to maintain their spin direction. The Earth's axis points at 23.5° to the vertical, so, when the northern hemisphere is presented to the Sun (summer), the southern hemisphere tilts away (winter), and vice versa.

As George Hadley realized in 1735, hot air rising at the equator should travel to the poles, where it cools and descends. However, such a simple "Hadley cell" circulation occurs only on a nonrotating planet. On Earth, air traveling poleward from the fast-spinning equator finds itself moving faster than the ground below. From the surface, it appears deflected eastward. This "Coriolis force" explains the direction of trade winds near the equator. Actually (and isn't this always so?) things are still more complicated. Rather than there being a single Hadley cell in each hemisphere, the circulation in fact consists of three overturning cells, or bands, of air.

The Coriolis force also explains why air flows anticlockwise around low-pressure (cyclonic) systems in the northern hemisphere and clockwise around them in the southern hemisphere. It is not true, as is popularly believed, that water swirls down a plug hole differently in the north and south. This is a phenomenon confined to large expanses of air.

The Earth's atmosphere is made up mostly of nitrogen–and oxygen, the by-product of life.

▲ Earth's clouds and weather are limited to the atmosphere's lowest layer—the troposphere, rising no more than 20 kilometers above the surface. The scattering of blue light by atmospheric gases continues into the stratosphere, up to 50 kilometers high. Above that, in the mesosphere, Earth's blue halo gives way to the blackness of space by about 100 kilometers above the surface.

▼ A line of thunderstorms builds over the rain forest as the afternoon sun glints on the waters of the Rio Madeira in the Amazon Basin of South America.

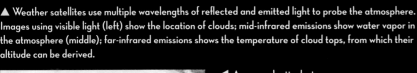

▲ Weather satellites use multiple wavelengths of reflected and emitted light to probe the atmosphere. Images using visible light (left) show the location of clouds; mid-infrared emissions show water vapor in the atmosphere (middle); far-infrared emissions shows the temperature of cloud tops, from which their altitude can be derived.

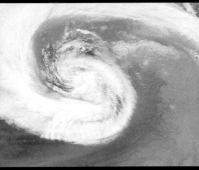

◀ An apocalyptic dust storm left daytime skies as dark as midnight across northern China in April 2001. The sand and dust was caught up in a powerful cyclone that carried it huge distances, even reaching across the Pacific to the Great Lakes of North America.

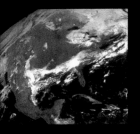

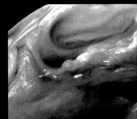

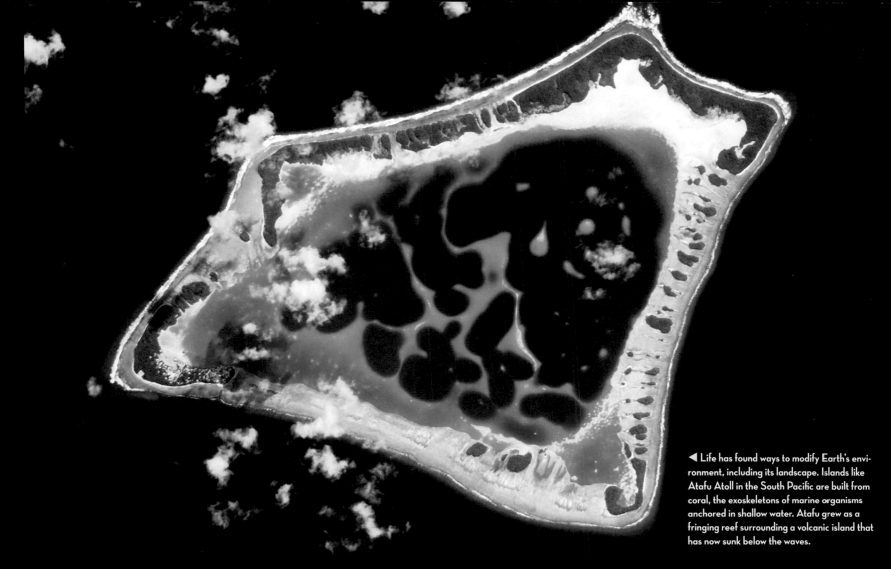

◄ Life has found ways to modify Earth's environment, including its landscape. Islands like Atafu Atoll in the South Pacific are built from coral, the exoskeletons of marine organisms anchored in shallow water. Atafu grew as a fringing reef surrounding a volcanic island that has now sunk below the waves.

Living planet

EARTH IS the only place in the universe where life is known to exist. With only one example, life is difficult to define. However, life's central characteristics include the ability to reproduce, move around, compete for resources, and pass information between generations.

All life on Earth uses common biochemical machinery based on a molecule called deoxyribonucleic acid, or DNA. This is a template for the building of proteins, "Swiss Army knife" molecules that do everything from making the scaffolding of cells to transporting oxygen in the blood to responding to light falling on the eye. The common working of all living things suggests a common origin.

Charles Darwin's genius was to realize that all living things have evolved from a simple ancient ancestor. Those with the traits necessary to survive and produce the most offspring have

proliferated. By this process of "natural selection," organisms have gradually morphed, creating a bewildering array of species.

According to fossil evidence, life arose almost immediately as the Earth was cool enough, about 3.8 billion years ago. In the laboratory, however, scientists have failed to create life from nonlife, implying the step is hard. One controversial explanation is that Earth was seeded by ready-made microorganisms, carried inside impacting comets.

Brandon Carter, ex-office mate of Stephen Hawking, provided an ingenious mathematical argument that there were five low-probability, "hard," steps on the road to humans. Step 1 was the advent of the first bacteria; step 2, complex cells with nuclei; step 3, multicellular life; step 4, intelligence; and step 5, human civilization. Each step took roughly 800 million years.

▲ Apart from the green of plants, there are other signs that Earth is a planet burgeoning with life. One is the red color of Lake Natron in the Great Rift Valley of East Africa. The color is from salt–loving microorganisms that thrive in the alkaline waters of this soda lake, which are hostile to most other life-forms. These algae pass their pigment on to the lake's more famous residents, pink flamingoes.

Earth's umbrella

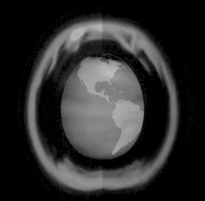

IMAGINE IF THE average temperature of the Earth were -18°C., the temperature our planet would have without the greenhouse effect. It gets universal bad press but the truth is, without it, we would not be here.

The key thing to realize is that visible light from the Sun is hardly absorbed by the atmosphere. It travels right through, which is why we can see the Sun. However, sunlight is absorbed by the ground, warming it. The ground then radiates the heat as invisible, "far infrared," light. Crucially, this is absorbed by greenhouse gases in the atmosphere. The principal one–water vapor–is the reason you are not freezing to death.

Carbon dioxide is the greenhouse gas whose levels we are boosting by burning fossil fuels like oil and coal. Historical records show the global temperature increasing since the start of the industrial age, in step with carbon dioxide in the atmosphere. Changes in the brightness of the Sun are not to blame.

The greenhouse effect is not the only "umbrella" that protects life on the Earth's surface from the harshness of space. We are also protected from deadly particle radiation from the Sun by the shield of the Earth's magnetic field. In addition, a high-altitude layer of ozone gas (ironically, a health hazard at ground level in cities) protects us from dangerous solar ultraviolet light. Without this ozone layer–composed of an unstable form of molecular oxygen–life could survive only in the seas.

▲ Earth's shield of stratospheric ozone protects the planet's life from ultraviolet solar radiation, which damages DNA. A cross–section through the Earth's atmosphere shows ozone concentration in January (left) and in October (right), when levels drop significantly over Antarctica in the southern spring.

How do we know the Earth is round?

IT IS NOT obvious the Earth is round. Apart from wrinkles like mountains, it seems flat. The Earth is big and its curvature so slight that we tend not to notice. But this is not always the case.

Ships at sea disappear over the horizon before they dwindle to a speck. On a flat Earth that would not be so. During a lunar eclipse, when the Earth passes between the Sun and Moon, the Earth's shadow on the Moon is curved. If people sail far enough in one direction, eventually they return to their starting point. And, of course, there are photos of Earth from space. Even the distances between any four cities have a mutual relationship that would be different if the Earth were flat, not round. In *Gravitation and Cosmology*, Steven Weinberg even uses such relations to deduce the curvature of Tolkien's Middle-earth.

In around 240 BC Eratosthenes made the first estimate of the Earth's size. He noticed that, at noon on the summer solstice, a vertical pillar at Syene (Aswan) cast no shadow, because the Sun was overhead, whereas the shadow of a pillar at Alexandra showed the Sun about 7° from the vertical. Knowing the separation of the locations and that 7° is about one-fiftieth of a circle, Eratosthenes calculated the Earth's circumference, and hence its diameter. His 7,800 miles is only 100 miles short of today's value.

Actually, the Earth is not a perfect sphere. At the equator, the ground is rotating at about 1,700 kilometers per hour, causing the planet's waistline to bulge. Unevenness in the planet's interior also makes the average level of the crust undulate, creating a knobbly figure called a "geoid."

▲ Eratosthenes, in the third century BC, determined the world's circumference to within a whisker of its correct value.

► "World Island" is one of a series of large-scale land reclamation projects off the coast of Dubai in the Persian Gulf. The largest man-made islands on Earth, the World, the Palm Jumeirah, and the Palm Jebel Ali give the city-state an extra 520 km of beach-front.

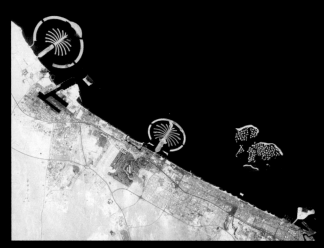

► The Great Plains of North America are a semi-arid area where irrigation is often required to grow crops. The circular irrigated fields in this area in southwest Kansas are 800 or 1600 meters across. The green fields contain mature crops; the brighter fields are recently ploughed or planted.

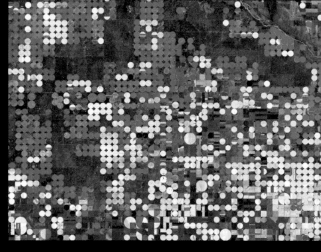

► The streets and buildings of Manhattan mostly follow a regular grid of city blocks, framing the green rectangle of Central Park. One exception is the diagonal path of Broadway, which follows an ancient Native American track.

► Once completely covered by dark green natural forest, a large part of the Bolivian state of Santa Cruz now shows the light green and brown stripes of agricultural development.

▼ At night over Egypt, light shines out from settlements along the River Nile. The cities of Cairo, Alexandria, Tel Aviv, Amman, and Damascus can be seen. Also visible in this view is the faint yellow band of "airglow" where atmospheric atoms and molecules interact with solar radiation at an altitude of 100 kilometers.

▼ The International Space Station is mankind's permanent presence in space. The station is about the size of a football field, and offers more living space than the average five-bedroom house. Orbiting at an altitude of 354 kilometers, the crew of six astronauts have great views of the planet below.

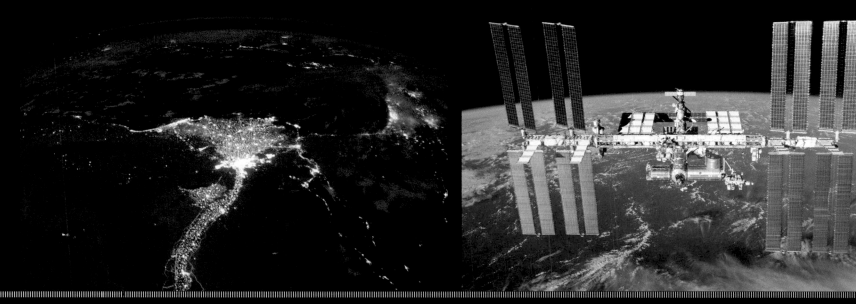

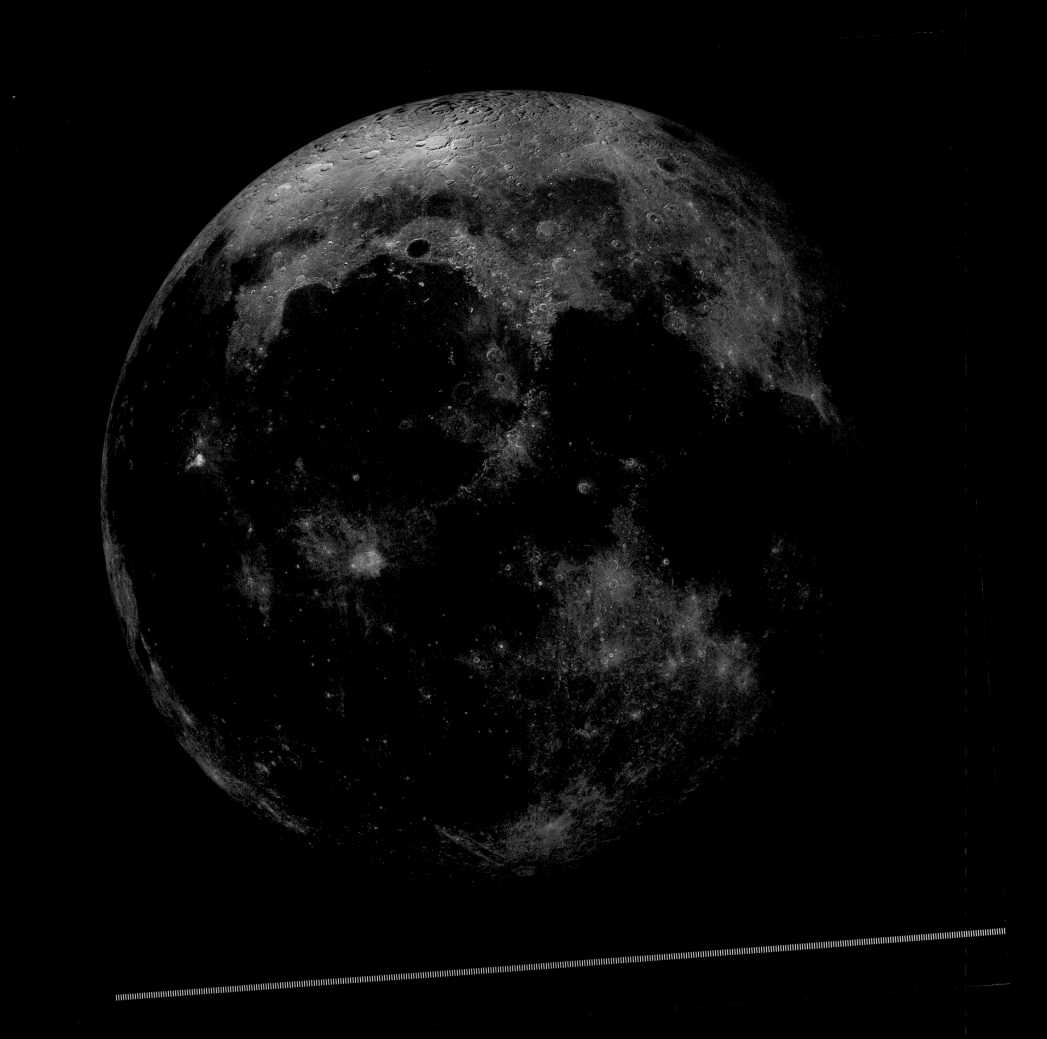

The Moon

THE MOON IS, by a very large margin, the closest celestial body to the Earth. It is the only body that appears as a true world to the naked eye. Millions of years ago on an African plain, even our remote human ancestors must have looked up and wondered what it is. In the pre-industrial age the Moon was a beacon that lit the way for travelers in the depths of night. Later, in the age of science, we learned that its importance is far greater: it pulls the tides in the oceans and even stabilizes the Earth's climate, keeping conditions on our planet favorable for life. But the true miracle of the Moon remains its proximity. The Moon is the only other body in the Solar System that has been visited by humans.

ORBITAL DATA

Distance from Earth 363,000 to 406,000 km
Orbital Period (Year) 27.28 Earth days
Length of Day 27.32 Earth days
Orbital Speed 1.1 to 1.0 km/s
Orbital Eccentricity 0.0549
Orbital Incination 18.3°
Axial Tilt 6.68°

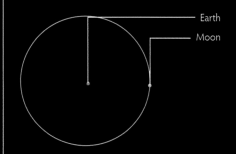

Earth
Moon

PHYSICAL DATA

Diameter 3,476 km / 0.27 x Earth
Mass 74 billion billion metric tons / 0.01 x Earth
Volume 22,000 million km^3 / 0.02 x Earth
Gravity 0.166 x Earth
Escape Velocity 2.375 km/s
Surface Temperature 40° to 396°K / −233° to 123°C
Mean Density 3.340 g/cm^3

Australia

ATMOSPHERIC COMPOSITION

Helium 50%
Argon 50%

Rocky crust
Silicate upper mantle
Transition zone
Solid iron core

Surface temperature

800 K
600 K
400 K
200 K
0 K

400°C
200°C
100°C
0°C

Mean density

0
1g/cm^3
2g/cm^3
3g/cm^3
4g/cm^3
5g/cm^3
6g/cm^3
7g/cm^3

Iron
Rock
Water

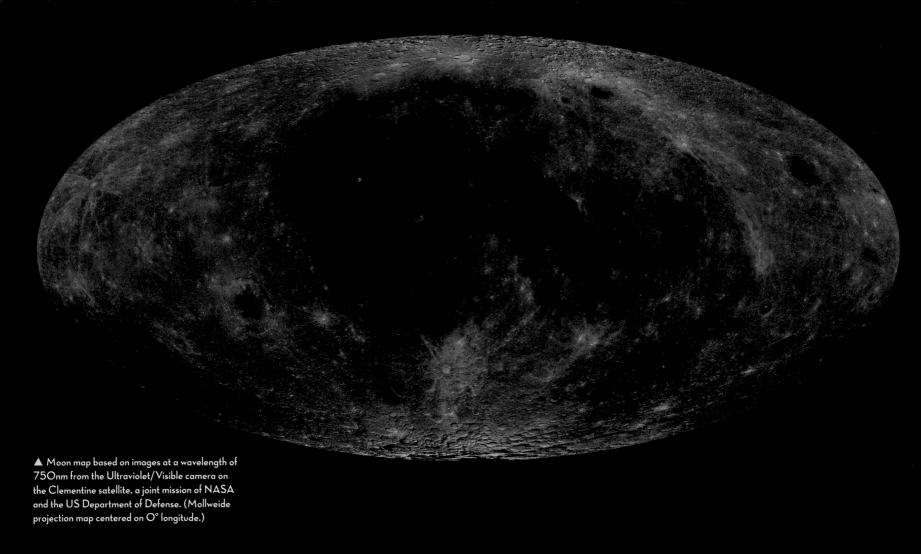

▲ Moon map based on images at a wavelength of 750nm from the Ultraviolet/Visible camera on the Clementine satellite, a joint mission of NASA and the US Department of Defense. (Mollweide projection map centered on 0° longitude.)

◄ A color view of the Moon taken by the Galileo probe as it left Earth for Jupiter in 1992.

◄ The circular impact basin Orientale lies between the bright highlands of the Moon's far side to the left, and the darker plains (mare) of the near side to the right.

◀ A crescent Moon, seen from the International Space Station in low Earth orbit in September 2010.

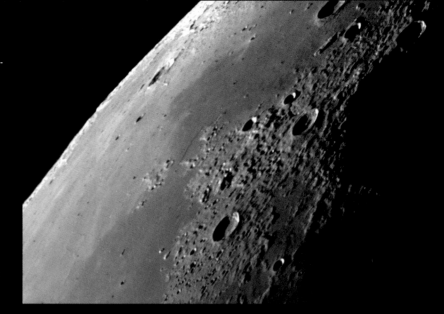

▶ Shadows creep across the circular basin of Sinus Iridium in this ground-based telescope view.

◀ An oblique view from lunar orbit looking towards Rima Ariadaeus, a 300 kilometer-long rift valley.

◀ This false color image shows variations in the chemical composition of the area surrounding the 42-kilometer-diameter crater Aristarchus, in the Oceanus Procellarum.

◀ The close-up view appears to show the track left by a large boulder (about 10 meters across) as it rolled downhill, finally coming to rest in a small crater.

▲ The Moon's south polar region is a potential site for a scientific base. The crater Shackleton (foreground) is permanently shadowed and so may contain water ice, while the nearby Malapert Mountain (background) is high enough to site solar power and communications equipment.

The men in the Moon

TWELVE MEN HAVE walked on the Moon, and one is buried there: Gene Shoemaker. Famous for proving that Arizona's giant Meteor Crater is an impact scar, Shoemaker not only trained the Apollo astronauts to recognize rocks but persistently lobbied NASA to send a scientist to the Moon. The American space agency, focused more on the race to beat the Russians, finally acquiesced, and geologist Harrison Schmitt flew on the last mission, Apollo 17. In recognition of Shoemaker's role, NASA's Lunar Prospector spacecraft, which was crashed onto the Moon in 1998, carried the geologist's ashes. He remains the only person buried–or at least scattered– on the Moon.

Of the men who stepped on the Moon between July 1969 and December 1972, the first two–Apollo 11's Neil Armstrong and Buzz Aldrin–came closest to disaster, touching down with just 45 seconds of fuel left. What Armstrong and Aldrin and their successors beheld was a truly alien world: a gray, crater-strewn desolation beneath an inky black sky.

Being small, the Moon's gravity is only one–sixth of the Earth's and its horizon a claustrophobic 2.5 kilometers away. With no atmosphere to scatter light, there is a huge contrast between bright areas and shadows, making photography difficult (incredibly, there is no photograph of the first man on the Moon–Aldrin missed the opportunity). There is also no atmospheric blurring of distant objects, which on Earth reveals what is near and far. Consequently, the Moon is a surreal world where there is no distinction between a 20-meter hillock 20 meters away and a two-kilometer mountain two kilometers away. And everything is coated in filthy, choking, clinging moondust.

◀ The impression of Buzz Aldrin's boot left in the lunar soil during the first manned landing on the Moon in July 1969.

▶ Gene Shoemaker, posing here at Arizona's meteor crater, is buried on the moon.

▼ Lunar landscape with the Apollo 17 lunar rover.

◄ Astronaut Buzz Aldrin poses for a photo on the lunar surface. Neil Armstrong took the picture and can be seen reflected in Aldrin's visor.

► Astronaut Buzz Aldrin unpacks experiments from the cargo bay of the Apollo 11 Lunar Module "Eagle."

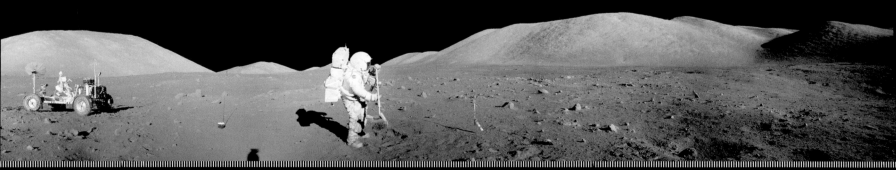

A fall of moondust

IN ARTHUR C. CLARKE'S *A Fall of Moondust*, the lunar dustcruiser Selene sinks with all its passengers into a sea of lunar dust. In 1961, when the novel was published, there was a very real fear that parts of the Moon were covered in a deep layer of quicksand-like dust. Although such fears proved unfounded, the Moon is indeed covered with a thin layer of fine dust, which does pose a potential hazard to future exploration.

The Apollo astronauts could not get the moondust off their spacesuits. It got into every nook and cranny of their spacecraft and smelled, they said, of gunpowder. Today, knowing that tiny "nanoparticles" can lodge in the lungs, causing breathing problems, there is a fear that moondust could be toxic. Certainly it could clog up the seals of spacecraft airlocks, causing a catastrophic malfunction.

Moondust is created when sand-grain-sized "micrometeorites" slam into the lunar surface, shattering and heating the rock. The resulting dust particles are like tiny melted snowflakes, quite different from smooth terrestrial sand grains. It is because they are so jagged that they snag on clothing. Their shape also causes them to reflect sunlight differently depending on their orientation, explaining the astronauts' observation that the lunar surface, far from being gray, shimmers with beautiful colors, from brown to gold to silver. The continual bombardment by micrometeorites turns over the lunar "soil" every few million years so, although the footprints left by astronauts will last a long time compared with on Earth, they will not last forever.

In addition to micrometeorites, the Moon is also of course bombarded by large meteorites.

▶ Three days on the lunar surface left Eugene Cernan of Apollo 17 caked in moondust.

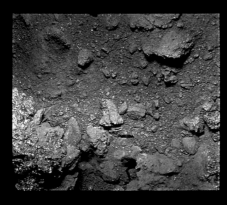

▲ A close-up view of an area 3 inches across reveals fine details of some of the rocks examined by the Apollo astronauts on the Moon's surface.

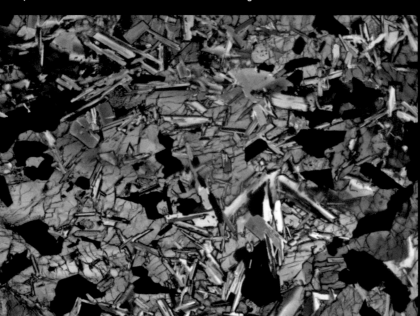

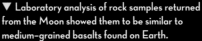

▼ Laboratory analysis of rock samples returned from the Moon showed them to be similar to medium-grained basalts found on Earth.

▼ Rocks returned by the Apollo astronauts allow scientists to examine samples of the early solar system, created 4.5 billion years ago.

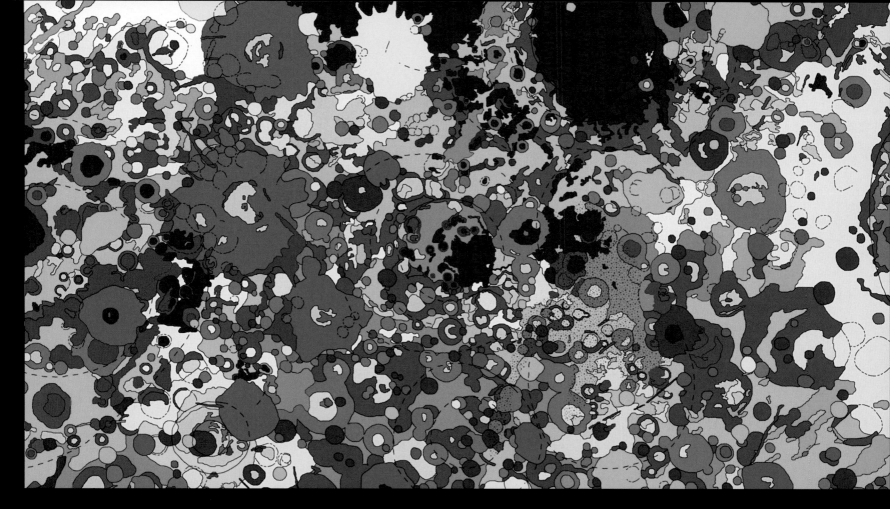

History book in the sky

▲ Jackson Pollock painting? No, it's a geological map of part of the Moon. Yellow, blue, and brown show impact craters of increasing age; red and pink reveal lava–filled basins, created by impacts 3.8 billion years ago.

ABOUT 800 MILLION years ago, an asteroid the size of Key West hit the Moon. It created the striking 93-kilometer–wide Copernicus crater and splattered debris far and wide. The Earth has come under similar bombardment; in fact, our planet is a bigger target. But the continual reworking of the Earth's surface by weather and the movement of plates has erased the evidence. Written in the Moon's battered face is the story of the history of the Earth.

Just as on Earth we find meteorites from the Moon and Mars, on the Moon we should discover meteorites from the Earth, ejected as debris from impacts. Intriguingly, lunar rocks may now be preserving biological material and even fossil microorganisms from the dawn of life on Earth–evidence long ago erased

by terrestrial geological activity. We may have to go to the Moon to find out about our origins.

The biggest impacts on the Moon and the Earth–far bigger than Copernicus–occurred 3.8 billion years ago. During this Late Heavy Bombardment, Jupiter and Saturn worked in concert to stir up the asteroid (or comet) belts and send bodies the size of Los Angeles our way. So big were the impacts that they punctured the lunar crust, causing lava to well up and flood the giant Mare basins, believed at one time to be lunar seas. Even now, cracks along the boundaries of the Mares occasionally erupt gas, seeping from the lunar interior as the Moon is alternately stretched and squeezed by tidal forces.

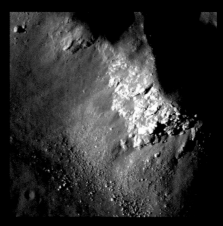

▲ Bright bedrock is exposed on the upper slopes of the central peak of the crater Copernicus.

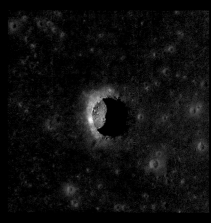

▲ A high sun angle illuminates the floor of an 80-meter-wide collapse pit in Mare Tranquillitatis (the Sea of Tranquility). A fast–flowing underground river of lava may have left a cave or tube, the roof of which collapsed after the lava drained.

The tides

TWICE A DAY the sea advances up the beach and retreats. Isaac Newton explained why. The Moon's gravity pulls strongest on the ocean facing it, less hard on the center of the Earth, and least hard on the ocean facing away. Consequently, the oceans bulge in two directions: on one side because the water is pulled away from the Earth, and on the other because the Earth is pulled away from the water.

The cause of tides is therefore not gravity but differences in gravity. As the Earth revolves every 24 hours, the tidal bulges travel around the Earth, creating two tides a day. There are tides in the Earth's rock too, though they are smaller because rock is stiffer. The Large Hadron Collider near Geneva expands and shrinks as the Moon alternately stretches and squeezes the 27-kilometer ring of the particle accelerator.

The Sun also creates tides, though a third as big. When Sun and Moon pull together, we get the highest tides. The Earth also pulls tides on the Moon–81 times as powerfully since the Earth is 81 times as massive. These cause moonquakes and occasional eruptions of gas. Over time, such tidal forces have braked the Moon's rotation so it essentially keeps one face to the Earth; it revolves on its axis in the same time, 28 days, that it circles the Earth. However, all sides of the moon receive sunlight: as the moon orbits the Earth different parts are illuminated by the Sun, resulting in the "phases" we see. There is no permanently dark side of the Moon.

Tidal movement is sapping the Earth of energy, causing it to spin ever more slowly and the Moon to recede from the Earth.

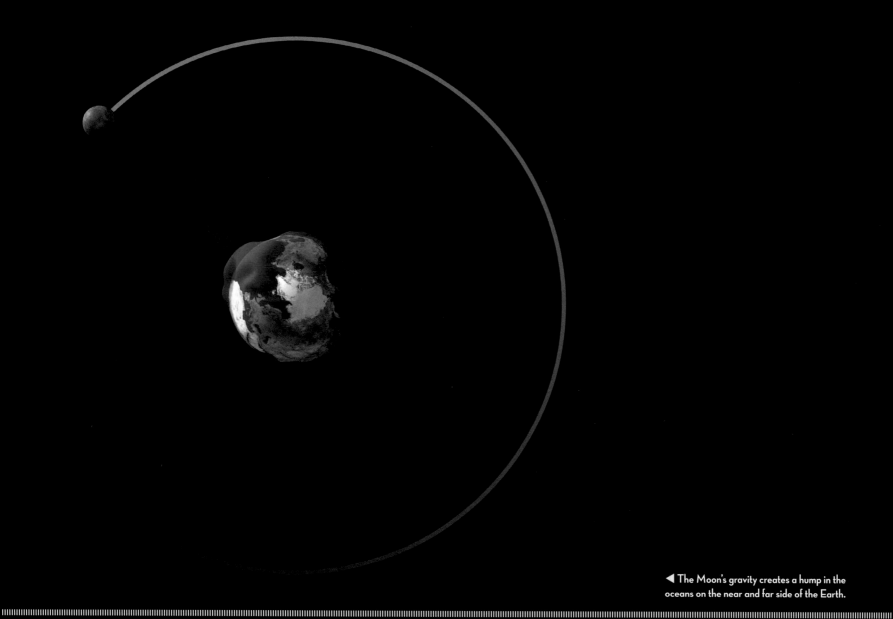

◀ The Moon's gravity creates a hump in the oceans on the near and far side of the Earth.

▼ The Laser Ranging Retro-reflector deployed during the Apollo 14 mission.

▼ A corner-cube reflects back light in exactly the direction it came from.

A tale of five corner-cubes

THE MOON IS moving away from the Earth at about 3.8 centimeters a year. How do we know so precisely? Because we can bounce laser light off reflectors left on the lunar surface by American and Russian spacecraft. The fist-sized reflectors, known as "corner-cubes," reflect light back in exactly the direction it comes from. From timing how long it takes light to return to the Earth, we calculate the distance to the Moon.

The corner-cubes left on the lunar surface fly in the face of the conspiracy theorists' claim that humans have never visited the Moon. They were left by the manned American spacecraft Apollo 11, 14, and 15, and the unmanned Russian rovers, Lunokhod 1 and 2.

The Lunokhod 2 reflector works

occasionally. The one on Lunokhod 1 was lost for almost 40 years, until the Lunar Reconnaissance Orbiter probe recently imaged the landing site. The coordinates were passed to scientists in New Mexico. On April 22, 2010 they fired a pulse of laser light at the landing site and were stunned to receive a return burst of 2,000 particles of light, or "photons." With four, and possibly five, corner-cubes now in action, it will be possible to observe not only the recession of the Moon but changes in its shape as it is tidally stretched and squeezed by the Earth.

The Moon's gradual recession from the Earth tells us something remarkable about our good luck in being born today. It involves eclipses.

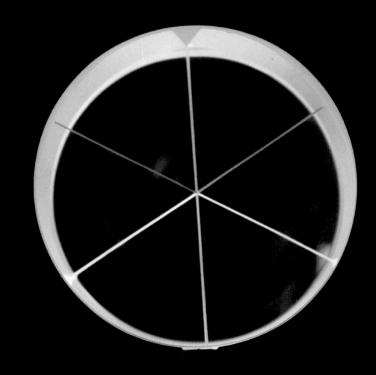

Total eclipse of the heart

DURING A TOTAL eclipse, the Moon passes between the Earth and the Sun, plunging the world into night in the middle of day. The temperature drops sharply, the wind gets up, bats start flying about, and the air is filled with the calls of frightened birds and animals. Staring up at the gaping black hole in the sky, prescientific people thought the Sun had been eaten by a monster, and banged pots and pans to frighten it away. (They always succeeded.)

A total eclipse is arguably the most spectacular of natural phenomena. But it is possible only because of a cosmic coincidence. The Sun, whose diameter is about 400 times that of the Moon, is also about 400 times further away. This means the two bodies appear the same size in the sky. Even though there are 170–odd moons in the Solar System, there is not another planetary surface from which such a perfect eclipse can be seen.

Actually, we are even luckier than this. Because the Moon is moving away from the Earth, in the past it appeared bigger and in the future it will appear smaller. Total eclipses of the Sun have been visible for only about 5 percent of the history of the Earth. We are indeed very fortunate to be around to see them. And this is not the end of our good fortune.

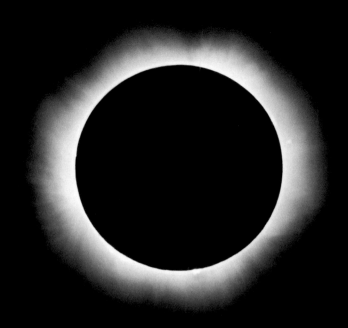

▲ Photo of the Sun's corona taken during an eclipse on August 11, 1999 that was visible across much of central Europe.

The planet that stalked the Earth

WHEN WORLDS COLLIDE, another world is born. Picture the Earth, shortly after its birth. A second world, the size of Mars, is bearing down on it. When the impact comes, it is so violent that the exterior of the Earth turns molten and is splashed off into space. It forms a ring–a ring around the Earth! This debris gradually congeals into a brand-new body. At first, the Moon is 10 times closer than it is today, raising tides 1,000 times higher than today in any water. But gradually, over billions of years, it moves out to its present position.

Is this how the Moon was born? Many are convinced it is. The key evidence came from the Apollo program. It found that the Moon is made of material suspiciously like the Earth's mantle, but drier than the driest terrestrial rocks, as if all their water was once driven out by intense heat.

There is a problem: For a Mars-mass object to create the Moon but not shatter the Earth, it must have made a glancing blow at a very low velocity. Bodies both inside and outside the Earth's orbit are moving far too fast.

The "Big Splash" theory can be made to work, however, if the Mars–mass body, called Theia, actually shared the same orbit as the Earth. This could have happened if it formed at a stable "Lagrange point," either 60° behind or 60° ahead of the Earth in its orbit. For millions of years, before being nudged into a colliding orbit, Theia bided its time. It was the planet that stalked the Earth.

1. Our satellite is believed to have formed from the molten rock splashed into space when the Earth was struck by a Mars-mass body. A body about the size of Mars formed on a similar orbit to that of the young Earth.

Thank your lucky stars

WITHOUT THE MOON, would we be here? Almost certainly not.

The Moon is unusually big, far bigger in comparison with its planet than any other satellite. Together the Earth and the Moon are essentially a "double planet.' The strong gravity of such a big Moon stabilizes the Earth's spin. If the planet threatens to tip over as spinning tops will do—the Moon pulls it upright again. Since such wobbles vary the sunlight reaching the ground, the Moon stabilizes our climate. Mars, which has no big moon, suffers catastrophic climate change. Life on Earth could never have evolved without a stable climate over billions of years.

Our big Moon also pulls large tides, which twice a day leave large tracts of the ocean margins high and dry. This drove the colonization of the land: stranded fish evolved lungs.

Our big moon has even driven science. By blotting out sunlight, total eclipses make it possible to see stars close to the Sun. In 1919 this permitted the observation of the bending of starlight by the gravity of the Sun, something predicted by Einstein's theory of gravity. Isaac Asimov, in his 1972 essay "The Tragedy of the Moon," even claimed that, if Venus had the Moon instead of the Earth, science would have arisen 1,000 years earlier. Asimov argued that, if people had seen Venus orbited by a visible moon, the "geocentric" idea of the Earth as the center of creation would never have been viable, and the Church could not have suppressed the ideas of those who thought otherwise.

But why do we have such a big Moon? The answer has to do with its unusual origin.

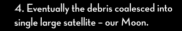
▲ As the astronauts on Apollo 8 completed their circumnavigation of the Moon on Christmas Eve 1968, they were greeted by the sight of the Earth slowly rising above the Moon's surface.

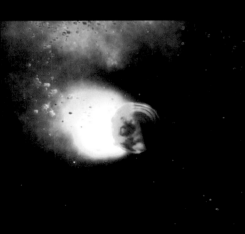

2. The two bodies collided, melting the smaller one and the Earth's crust, which was ejected out into space.

3. Some of the ejected globules rained back down on the Earth's surface, but some remained orbiting in a ring of debris.

4. Eventually the debris coalesced into single large satellite – our Moon.

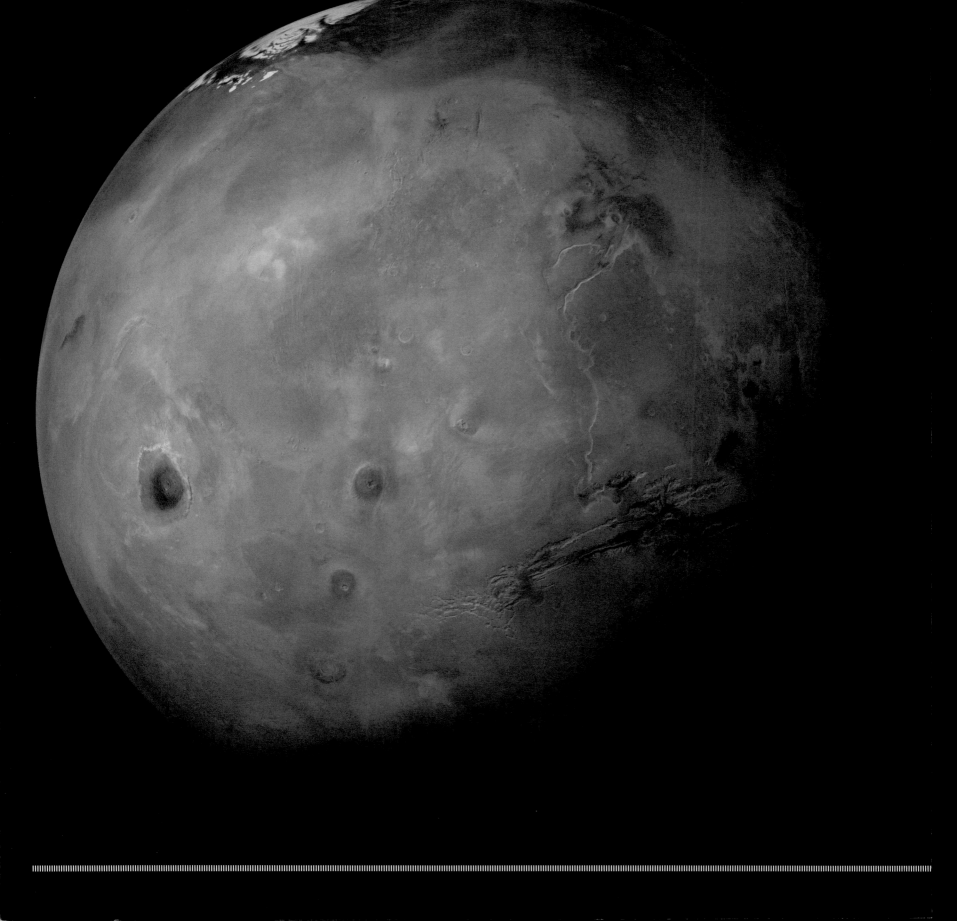

Mars

THROUGHOUT HISTORY Mars has beckoned, a brilliant ruby light in the night sky. In the space age we have heeded the call. Wave after wave of our robot emissaries have crossed the gulf to the Red Planet. As everyone knows, they are preparing the way for flesh and blood. Mars is the next frontier, a living, breathing world that humans expect to one day call home.

The Red Planet has an ultrathin atmosphere, is raked by deadly subatomic particles from the Sun, and barely reaches 0°C on a hot summer's day. Although a harsh environment, Mars is not dead. It is dynamic, with ice caps, giant volcanoes, clouds, and planetwide dust storms. Significantly, there is evidence of ancient rivers and possible oceans. And water raises the prospect of life—simple microorganisms, mind you; not an advanced Martian civilization.

ORBITAL DATA

Distance from Sun 206 to 249 million km / 1.38 to 1.66 AU
Orbital Period (Year) 686.78 Earth days
Length of Day 24.62 Earth hours
Orbital Speed 26.5 to 22.0 km/s
Orbital Eccentricity 0.094
Orbital Incination 1.85°
Axial Tilt 25.19°

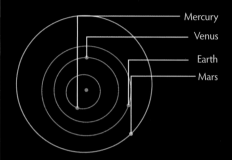

Mercury
Venus
Earth
Mars

PHYSICAL DATA

Diameter 6,794 km / 0.53 x Earth
Mass 642 billion billion metric tons / 0.11 x Earth
Volume 163,000 million km³ / 0.15 x Earth
Gravity 0.379 x Earth
Escape Velocity 5.022 km/s
Surface Temperature 133° to 293°K / −140° to 20°C
Mean Density 3.94 g/cm³

The Moon

ATMOSPHERIC COMPOSITION

Carbon Dioxide 95.3%
Nitrogen 2.7%
Argon 1.6%

Thin carbon-dioxide atmosphere

Rocky crust

Silicate mantle

Iron core

▲ Mars map based on selected images from the Viking 1 and 2 Orbiters. (Mollweide projection map centered on 0° longitude.)

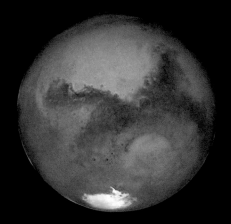

▲ Mars was at its closest to Earth for almost 60,000 years in August 2003, when it was briefly the brightest planet in the night sky.

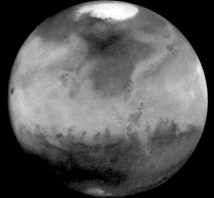

▲ Clouds form in the atmosphere of Mars as water ice from the north pole sublimates into the atmosphere during the northern summer.

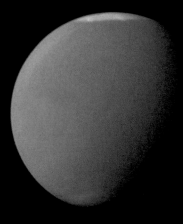

▲ A global dust storm obscured the surface of Mars for three months in 2001. Dust storms arise every northern springtime, but this one was unusually early and the largest for several decades.

▲ Not trees growing up, but sand falling down from the crests of sand dunes in the far north of Mars. The sand is dislodged as a frost of carbon dioxide ice evaporates in the spring, leaving dark streaks as it cascades down the sides of the dunes.

◀ A dune field on the floor of Endurance Crater, as seen by the Mars rover Opportunity. The slightly blue floor indicates rocks rich in haematite.

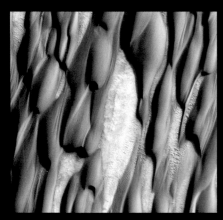

▲ Wind-blown dunes of blueish basaltic sand cover the Abalos Undae region. The dunes are overlaid by a more recent layer of red dust.

▼ Dark trails mark the paths taken by short-lived dust devils across the Martian plains. The bright surface layer is disturbed revealing darker material underneath.

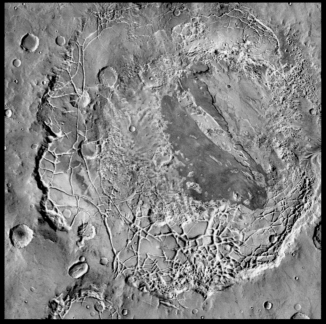

◀ Aram Chaos is an area of "chaotic" terrain, probably the result of ground ice melting, causing collapse of the surface and a sudden outflow of water to the east. Water-related minerals such as haematite and sulphates have been detected from orbiting spacecraft.

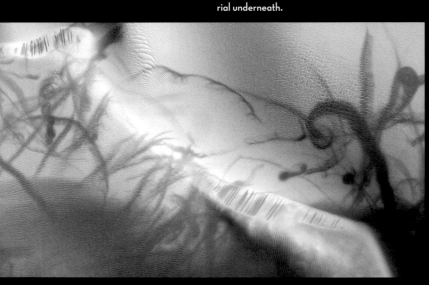

◀ Victoria Crater, an 800-meter-diameter impact crater in Meridiani Planum, explored by the Mars rover Opportunity from September 2006 to August 2008.

Mars in the imagination

ONCE UPON A TIME it was widely believed that Mars was home to a dying extraterrestrial civilization. The belief was triggered by Giovanni Schiaparelli, who observed the planet from Milan during its close approach to Earth in 1877. The Italian astronomer was convinced that his telescope showed a vast planetwide network of channels etched into the surface. The trouble was that the Italian for channels (*canali*) was very close to the English word for artificial waterways: "canals."

From his private observatory in Flagstaff, Arizona, Percival Lowell built on the work of Schiaparelli, sketching an intricate web of canals, too straight, he argued, to be natural. Lowell concocted a compelling story. Mars, in the grip of catastrophic climate change, was drying out. A great and noble civilization faced oblivion. In desperation, it had embarked on the engineering project to end all engineering projects: a gargantuan system of canals thousands of kilometers long down which water flowed to the parched equatorial regions from the Martian polar ice caps.

As images from space probes would later show, Schiaparelli and Lowell had been engaged in a game of connect-the-dots with Martian features teetering on the edge of telescopic visibility. The lines they drew were undoubtedly produced by intelligence. The question was: at which end of a telescope?

Mars has always stirred the imagination, ever since the Romans named it for their god of war because its color reminded them of blood on the battlefield. But it was Lowell's romantic view of Mars that stoked literary imaginations. Writers from H. G. Wells to Arthur C. Clarke to Kim Stanley Robinson have portrayed Mars in their work.

▼ Percival Lowell expanded on the work of Giovanni Schiaparelli, imagining a vast network of canals on the Red Planet. Here are some of his drawings.

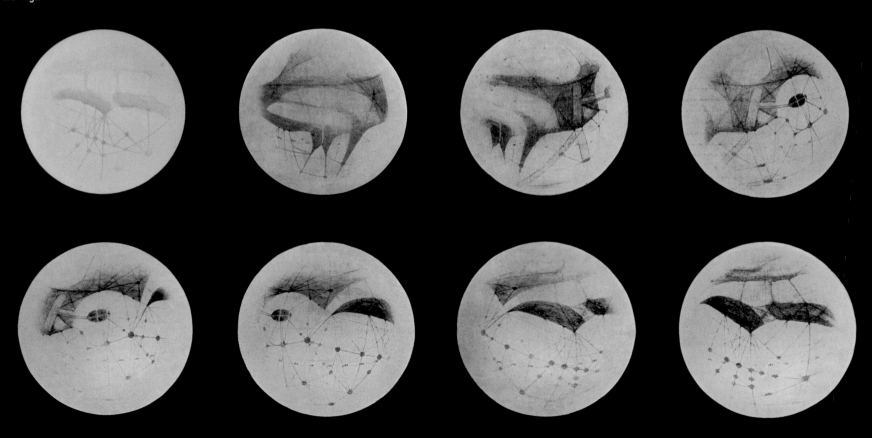

The real Mars

WHEN THE FIRST SPACE probes to Mars–NASA's Mariner 6 and 7–flew past in 1969, there was disappointment. Although only a few black-and-white images were radioed home, they revealed a desolate, crater-strewn world. Hopes of finding anything interesting were therefore not high when, on November 14, 1971, Mariner 9 became the first spacecraft to enter into orbit around another planet.

Unfortunately, the space probe had arrived at the height of a planetwide dust storm. Flour-like Martian dust, ground by wind and continual meteoritic bombardment, is easily lifted high into the thin atmosphere by warm air rising when the planet is closest to the Sun. Once there, the dust falls back to the surface only slowly in the weak gravity. Occasionally, as in November 1971, the whole planet is shrouded in dust.

It was a blessing in disguise. As the dust settled, it revealed first the highest features, then the next highest, and so on, providing crucial 3-D information. First to poke through were the summits of four truly gargantuan volcanoes. One–Olympus Mons–is about three times the height of Everest. Next was a giant crack in the skin of the planet–now known as the Valles Marineris–a canyon snaking a third of the way around Mars with tributaries bigger than the Grand Canyon. Add to this sand dunes and sinuous channels, and, overnight, our image of Mars was transformed.

Mariners 6 and 7, it turned out, both had flown over the Martian southern hemisphere, which is like the Moon. The northern hemisphere is strikingly different, a geological wonderland carved by volcanism and, very likely, running water.

▲ Olympus Mons is the largest volcano in the Solar System, 600 kilometers across and rising 26,000 meters above the surface of Mars.

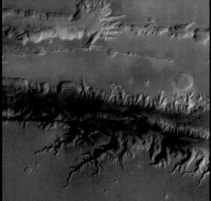

▲ Valles Marineris outstrips Earth's Grand Canyon on all counts: 3,000 kilometers long and on average 8,000 meters deep.

▶ The Tharsis ridge is a broad area of Mars that has been uplifted 10,000 meters and is topped by three immense shield volcanoes.

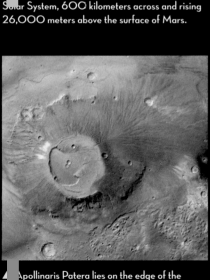

▲ Apollinaris Patera lies on the edge of the southern highlands of Mars. Its large summit crater, or caldera, is 60 kilometers in diameter.

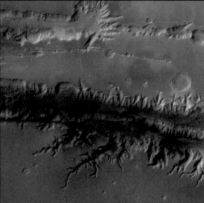

▲ Noachis Terra is a crater–strewn highland region in the southern hemisphere of Mars, seen here under a substantial atmospheric haze.

Water on Mars

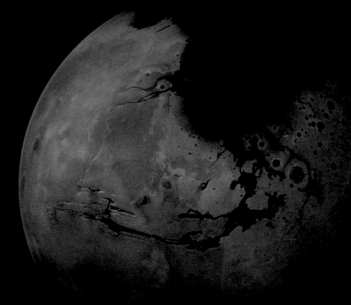

MARS BEARS the scars of water–river valleys and flooded plains and maybe even ancient oceans. But when did water flow on Mars? And where is it now?

Mars today is very different from ancient Mars. In its first billion or so years the planet had a much thicker atmosphere, pumped out from volcanoes. But it lost it. Why? Possibly it leaked into space because of the planet's weak gravity, or large impacts blew it away. Another reason could be that Mars, smaller than Earth, lost its internal heat more quickly. Its iron core solidified, stopping the circulating electrical currents that generate a planetary magnetic field. Bereft of this protective shield, the solar wind tore at its atmosphere.

Without an atmosphere, liquid water boils away and is lost. Liquid water may also have seeped into the porous, impact–shattered rock in the giant craters of Mars. The result may be an ice-rich layer at least a kilometer thick. Near the equator the ice could be about 400 meters down, judging by impact craters showing evidence of ejected icy slurry. In other places it could be only 100 meters deep. We know for certain that ice is still on the surface at the Martian poles.

The story of water on Mars is complex. Without the correcting effect of a large moon, Mars can tilt wildly on its axis. At times one hemisphere points almost at the Sun. Sunlight melts its ice cap completely, driving off carbon dioxide and water that then fall as snow or even rain in the cold hemisphere. During these brief warm periods, Mars may have flash floods and waterfalls and rainbows. And water, on Earth at least, is synonymous with life.

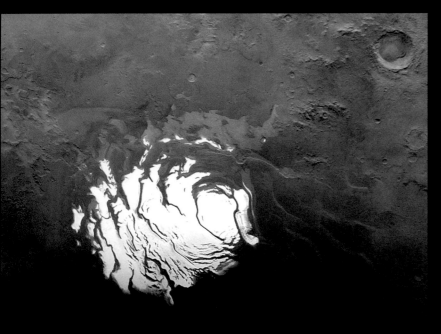

▲ The south pole of Mars is permanently covered by a small ice cap of frozen carbon dioxide, never less than 400 kilometers across. It is likely that layers of water ice lie beneath the carbon dioxide frost.

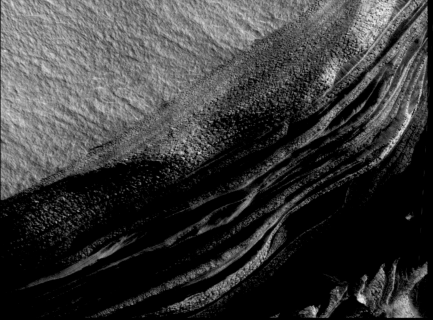

▲ The layered deposits around the north pole of Mars are alternating layers of water ice and wind–borne dust. This detailed image shows the ice–rich layers fracturing into blocks on the wall of a deep canyon.

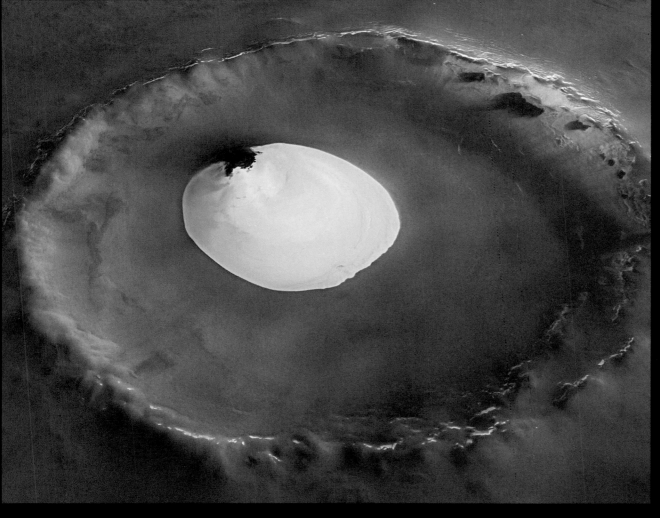

◄ Water ice survives the summer thaw at the bottom of this unnamed crater near the north pole of Mars, and in the shadows along its rim.

▼ Meandering and braided gullies emerge from the base of rocky cliffs at the rim of this southern-Martian crater.

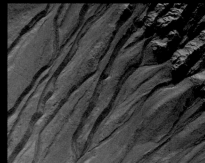

▼ The complex spiral of ice visible at the north pole of Mars is merely the top layer of a 2-kilometers–high stack of alternating ice and dust layers.

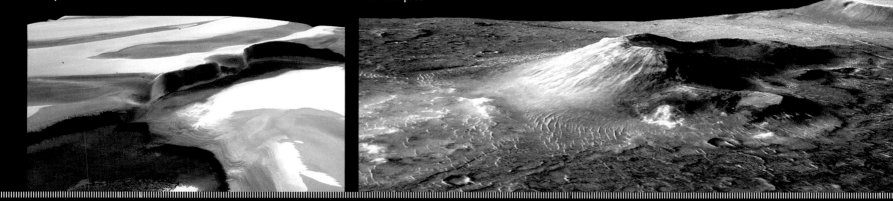

▼ Hydrothermal mineral deposits have been detected in infrared images of the volcanic cone Nili Patera. The deposits, showing as bright areas near the base of the cone, are evidence for warm, wet, or steamy conditions at some time in the past.

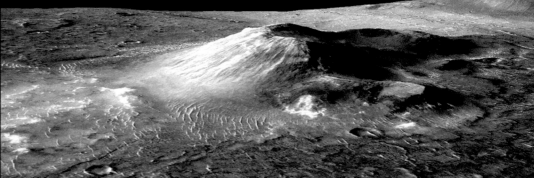

Exploring Mars

AROUND 50 SPACE PROBES have gone to Mars. The failure rate is high. Among the successes was Mariner 9. This mission, between 1971 and 1972, obtained the first complete map of the planet from orbit, revealing ancient river channels, mega-volcanoes, and fields of shifting sand dunes. Another milestone was Viking, whose two landers touched down in 1976. Although its biological experiments found no life, the mission did not dampen anyone's enthusiasm.

Interest in life on Mars was whipped up again in 1996. Scientists claimed that meteorite AH 84001, found in Antarctica's Allan Hills but originally from Mars, contained microscopic fossils of Martian bacteria. The claim remains controversial but it kick-started more Martian exploration.

In 2003, Mars Express revealed that there were three distinct Martian geological eras: an early one, the Noa-chian, before about 3 billion years ago, when Mars had a thick atmosphere and abundant water; a transitional period, the Hesperian; and the most recent, and longest, period, the Amazonian, with a drier, harsher climate, characterized by the rusty-red iron mineral that gives the Red Planet its distinctive color. Recent highlights have been Mars Phoenix in 2007, digging down and revealing white permafrost, and, from 2004 onward, two rovers–Spirit and Opportunity–finding sedimentary layers of rock that must have been laid down under ancient water.

Mars tantalizes us. The similarities add up. The sun rises every 24 hours, though diluted to roughly half its terrestrial strength. The planet's axial tilt is similar to Earth's, causing a similar pattern of seasons. Mars even has an atmosphere, though it is mostly carbon dioxide and a hundred times thinner than Earth's.

▲ Carbonate rocks were detected by the Mars rover Spirit on an outcrop called Comanche, in hills about 5 kilometers from its landing site. Carbonate rocks originate in wet, non–acidic environments that could be suitable for life.

▼ The tiny rover Sojourner investigates a rock in Ares Vallis, as viewed from its "mother ship," the Mars Pathfinder lander.

◀ The Mars rovers, Spirit and Opportunity, have a robotic arm carrying sensors for close–up examination of Martian rocks.

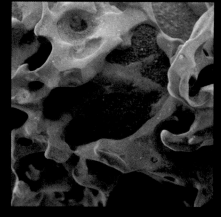

▼ Close-up examination of this rock, found on the surface of Mars, with a microscopic camera and an X-ray spectrometer confirmed it is a nickel-iron meteorite.

▲ Microscopic imaging reveals the jagged surface of a Martian lava flow. Small cavities left by gas bubbles have been revealed by billions of years of erosion by wind-blown sand.

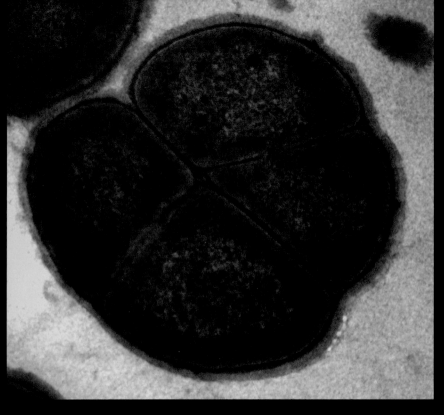

▲ The Sun sets over the rim of Gusev Crater at the end of the rover Spirit's 489th Martian day. Twilight can last up to two hours after sunset on Mars thanks to light scattered by high-altitude dust.

▲ A simulated view of the biggest volcano in the Solar System, based on height data and imagery from NASA's Viking and Mars Global Surveyor spacecraft.

King of the mountains

IMAGINE A VOLCANO three times the height of Everest and the area of Arizona. Imagine a volcano whose lava-producing crater, or "caldera," is roughly 70 kilometers across and bordered by cliffs up to three kilometers high. Such a mega-volcano exists–on Mars. It is called Olympus Mons, after the Greek home of the gods, and it pokes up through the thin Martian atmosphere, its summit shrouded in fog and dusted with snow.

On Earth, superhot lava wells up from deep in the planet but the moving tectonic plates ensure that such "blow torches" are never concentrated beneath the same patch of crust. Contrast this with Mars. With no tectonic activity, the blowtorch blisters one spot for billions of years, causing unending lava to spill onto the surface and build a truly enormous volcano like Olympus Mons.

Olympus Mons may be tall but,

like its smaller earthbound counterpart, Mauna Kea in Hawaii, its slopes are barely perceptible. Walking to the summit, you would climb only 40-odd meters every kilometer.

At 27 kilometers, Olympus Mons is the tallest mountain in the Solar System. But how does it compare with others? Next in height on Mars are the mega-volcanoes Ascraeus Mons (18.2 km), Arsia Mons (17.8 km), Pavonis Mons (14 km), and Elysium Mons (13.9 km). Venus boasts the volcano Maxwell Montes (11 km); Jupiter's moon Io, Boösaule Montes (17.5 km); and another of its moons, Iapetus, the extraordinary and mysterious Iapetus Ridge, which stretches 1,300 kilometers around the tiny world and in places is 20 kilometers high. Compare all these with Earth's anthills: Everest at 8.5 kilometers above sea level and Mauna Kea at 10.2 kilometers above the Pacific seabed.

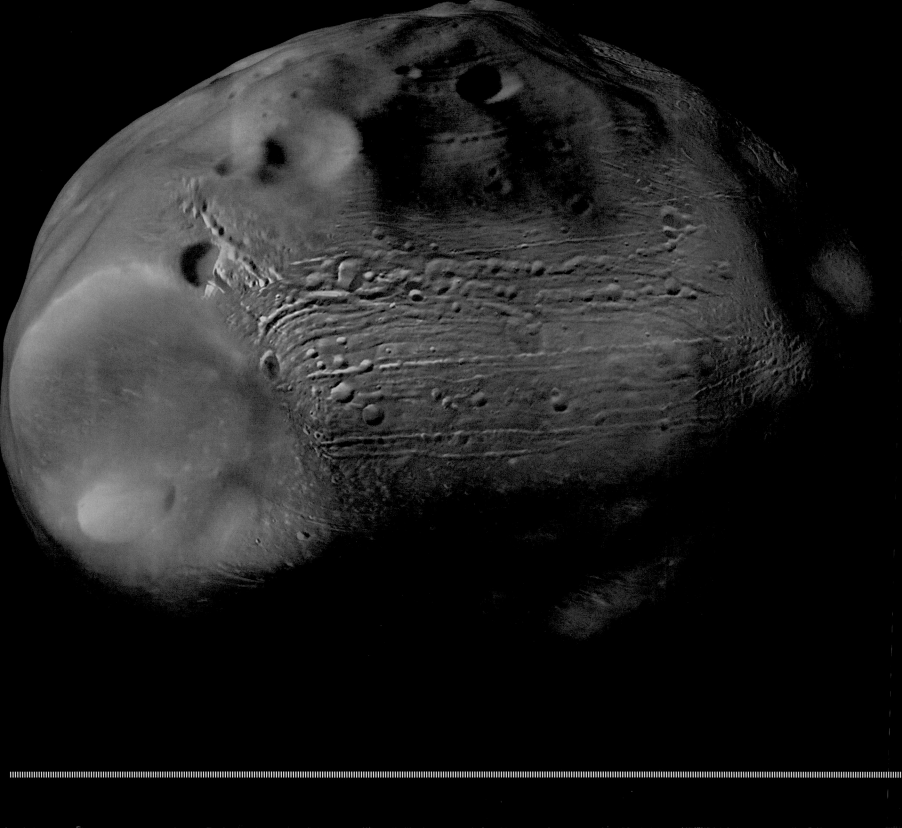

Phobos

ONE OF THE EXTRAORDINARY things about Mars's moons, Phobos and Deimos, is that their existence was anticipated in *Gulliver's Travels*. Jonathan Swift even got the moons' orbital periods almost correct. This was in 1726, 150 years before Mars's moons were actually discovered. Swift almost certainly knew of a prediction by 16th-century German astronomer Johannes Kepler who believed that geometric relations govern the heavens. Since the Earth had one moon and Jupiter four,

argued Kepler, surely the in-between planet, Mars, should have two?

Phobos and Deimos were discovered by U.S. Naval Observatory astronomer Asaph Hall in 1877. But he nearly missed out. After unsuccessful nights searching for Swift's moons, Hall gave up, despondent. However, his wife, Angelina Stickney, sent him right back to the telescope. He duly discovered and named Phobos (Fear) and Deimos (Terror) after the horses that pulled the chariot of the Greek god of war.

ORBITAL DATA

Distance from Mars 9,240 km
Orbital Period (Year) 0.32 Earth days
Length of day 0.32 Earth days
Orbital speed 2.2 to 2.1 km/s
Orbital Eccentricity 0.0151
Orbital Inclination 1.08°

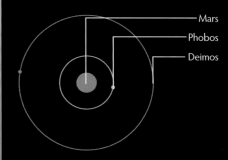

- Mars
- Phobos
- Deimos

PHYSICAL DATA

Diameter 22 km
Mass 11 million million metric tons
Volume 5,680 km³
Gravity 0.001 x Earth
Escape Velocity 0.011 km/s
Surface temperature 233°K / -40°C
Mean Density 1.75 g/cm³

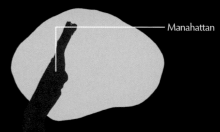

- Manahattan

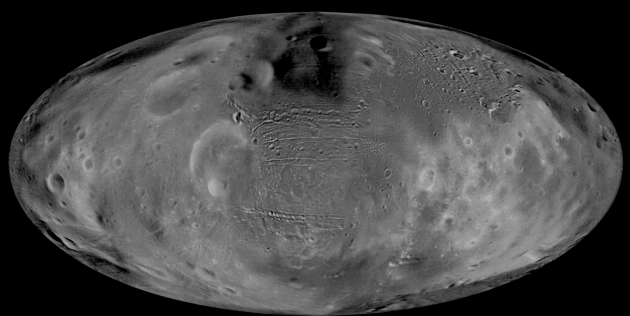

▲ Phobos map based on images from the European Space Agency's Mars Express satellite. (Mollweide projection map centered on 0° longitude.)

▼ Comparison of small Solar System bodies, which become more spherical with increasing size.

Mars' moon Phobos (26 km)

Asteroid Lutetia (132 km)

Saturn's moon Hyperion 370 km)

Neptune's moon Proteus (420 km)

Saturn's moon Encadeus (496 km)

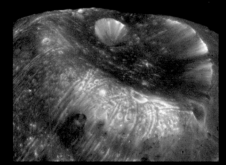

▲ The most prominent feature on Phobos is the crater Stickney, about 9 kilometers across, a significant proportion of the size of Phobos.

How Phobos came to be

THE BIG MOONS of the giant planets mostly congealed from debris disks surrounding their parent planets, much as the planets congealed from a disk around the Sun. But Phobos looks very similar to rocky bodies found in the Asteroid Belt between Mars and Jupiter–as in fact does Deimos, its companion. It's plausible, then, that these two, and many of the other small moons in the Solar System, are captured asteroids, bodies that strayed too close to planets and were grabbed by their gravity. Our Moon, which is believed to have "splashed" from the Earth in a titanic impact, is an example of a third way for a moon to be born.

The trouble with the captured-asteroid idea is that such moons tend to have highly random orbits–highly elliptical, oriented in any direction, with the moons sometimes going around the opposite way to the planet's spin. However, the orbits of Phobos and Deimos are not only circular but almost exactly in the plane of Mars's equator, as would be expected for moons that formed from a debris disk circling the planet.

There are other mysteries. The drag of the Martian atmosphere is causing Phobos to spiral into Mars. It was once believed that its rate of fall could be explained only if Phobos were hollow–an artificial alien artifact. With better orbital data, such a dramatic hypothesis is unnecessary. Nevertheless, the moon is unusually light, and may be like honeycomb. Phobos's fate, in about 40 million years, will be to slam into Mars. It's something to consider for any of our distant descendants who may then be on Mars and will see a moon fall from the sky.

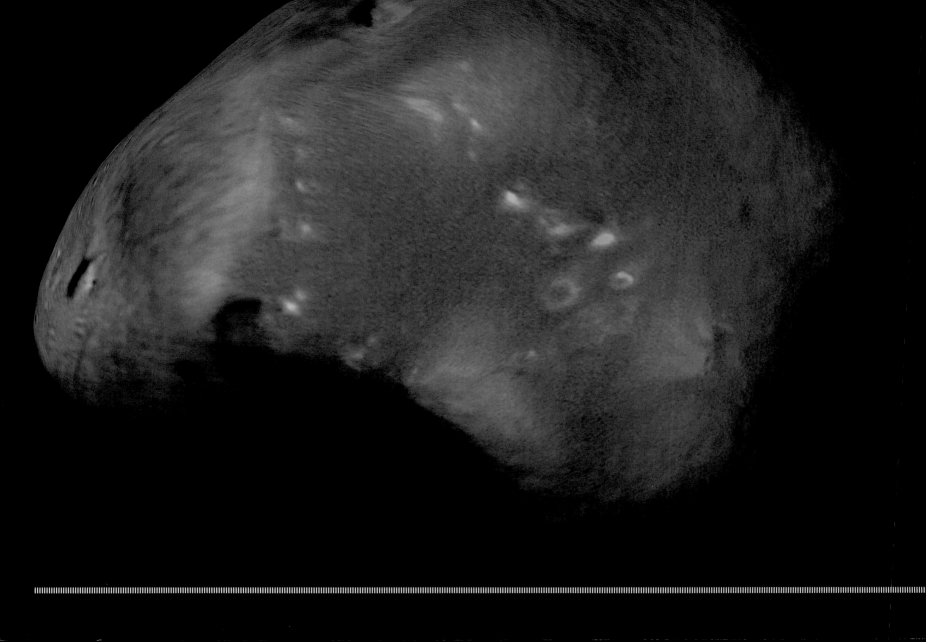

Could you jump off Deimos?

YOU ARE ON DEIMOS, space-suited and bored. You decide to entertain yourself by seeing how far you can long-jump. You sprint, you leap, and you sail across the coal-black, pockmarked terrain. Back on Earth you can jump a meter vertically. But here on Deimos, a moon with gravity a mere thousandth as strong, you go 1,000 times higher. Flying above the surface in a long, leisurely arc, you reach a height of a kilometer before, many minutes later, coming back down.

Since it was fun, with Mars dominating the sky, you try again, running faster before launch. This time you travel even further before coming down. You leap faster and go further still. Eventually, you fly so far that the surface of Deimos curves away below you as fast as you fall back toward it. Now you are falling for ever, in a circle. Unless you fire your rocket pack, you will never come back down. You have jumped into orbit.

This is what Earth satellites are doing. This is why they do not come down. They are forever falling but never reaching the ground. On Earth, this requires an "orbital velocity" of 7.8 kilometers per second, or 28,000 kilometers an hour. But on Deimos 3.75 meters per second, or 13.5 kilometers an hour, is sufficient. Fit, young people can achieve this, and launch themselves into orbit. Going about 40 percent faster, they can even attain "escape velocity." Imagine jumping off Deimos and falling down through the thin atmosphere of Mars. One day, interplanetary thrill-seekers may do just that.

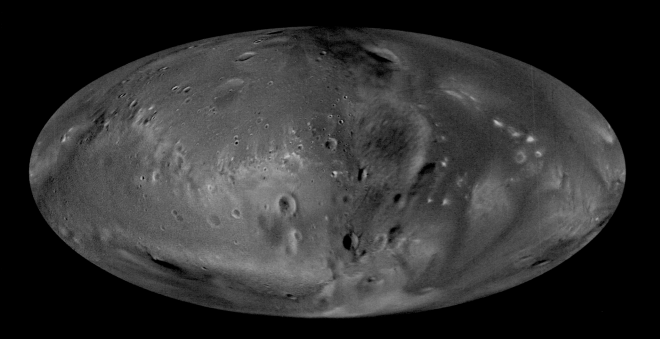

▲ Deimos map based on images from NASA's Viking Orbiter. (Mollweide projection map centered on 0° longitude.)

ORBITAL DATA
Distance from Mars 23,400 to 23,500 km
Orbital Period (Year) 1.26 Earth days
Length of day 1.26 Earth days
Orbital speed 1.35 km/s
Orbital Eccentricity 0.0005
Orbital Inclination 1.79°

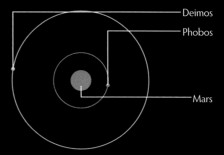

Deimos
Phobos
Mars

PHYSICAL DATA
Diameter 6 km
Mass 2 million million metric tons
Gravity 0.002 x Earth
Escape Velocity 0.01 km/s
Surface temperature 233°K / –40°C
Mean Density 1.90 g/cm³

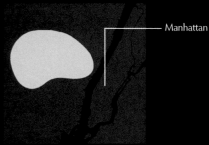

Manhattan

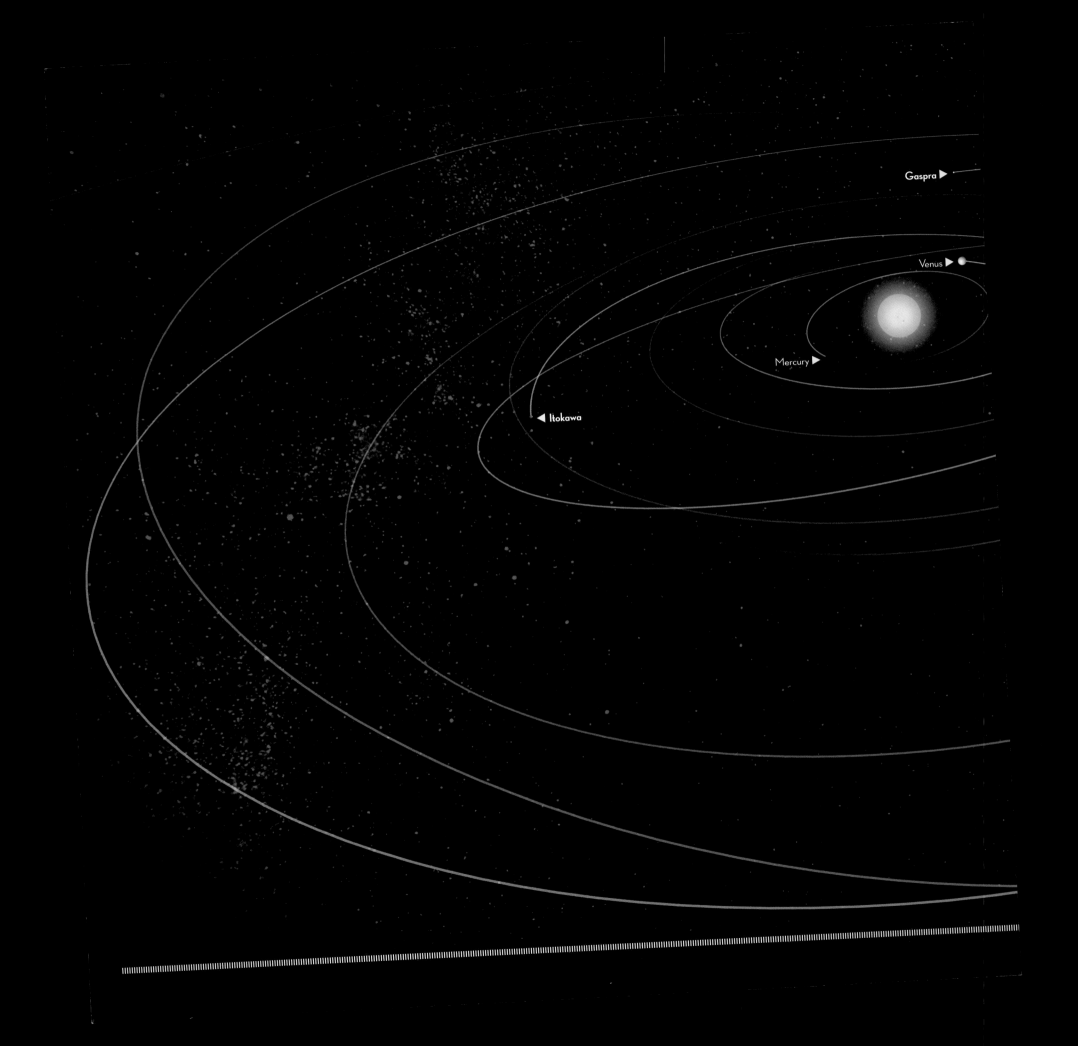

Jupiter ▶

Ceres ▶

Earth ▶

Eros ▶

◀ Mars

◀ Ida

Asteroid Belt

THE FIRST ASTEROID was discovered in 1801.
Astronomers knew of planets and moons, but here was
something entirely new under the Sun: thousands upon
thousands of chunks of rocky debris tumbling through
space as they circled the Sun between Mars and Jupiter.
What were they? The obvious answer was the sorry
remains of a planet that had disintegrated or blown up
or met some other cataclysmic end. But the total mass
in the Asteroid Belt turned out to be only about one-
thousandth of the Earth's–too small to make a planet.
No, the asteroids are not a planet that died but one that
was prevented from being born. The evidence for this is
actually carved into the Asteroid Belt itself.

PHYSICAL DATA
Total Mass 3 billion billion tons (estimated)
Material Carbonaceous, silicate, or metallic
Number of Asteroids 700,000 to 1.7 million (estimated)
Asteroids Over 200km IN
Diameter Over 200
Largest Objects Ceres, Vesta, Pallas, and Hygiea

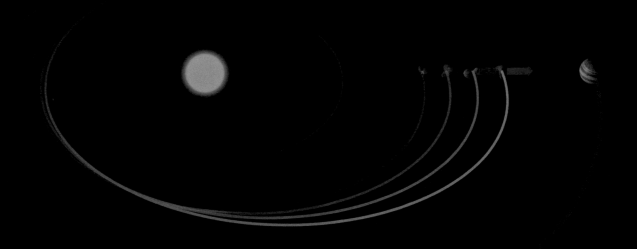

Gravitational no-go areas

BY 1857 the number of known asteroids had crept up to 50. Then an American mathematics professor noticed a pattern. Asteroids appeared to shun orbits at particular distances from the Sun.

Daniel Kirkwood guessed correctly that these gaps were carved out of the Asteroid Belt by the gravity of the nearby giant planet, Jupiter. It works like this. If an asteroid makes, say, two orbits of the Sun in the time it takes Jupiter to complete one, then periodically Jupiter and the asteroid pass close to each other on the same side of the Sun. As they do, Jupiter gives the asteroid a sharp tug. Over time, the effect of these tugs builds up—in exactly the same way pushing a swing regularly at the right rate makes it go ever higher. An asteroid in such a "resonant" orbit ends up being ejected from that orbit.

Prominent Kirkwood Gaps are known where the orbital period of Jupiter and asteroids have the ratios 4:1, 3:1, 5:2, 7:3, and 2:1. Rare asteroids have been found to orbit within the Kirkwood Gaps—for instance, asteroids of the Alinda family (in the 3:1 resonance) and the Griqua family (2:1). These bodies appear to avoid the disruptive effect of Jupiter because their orbits are not circular but highly elongated.

The result of "resonance" effects are seen all over the Solar System. Kirkwood also guessed correctly that a prominent gap in the rings of Saturn known as the Cassini Division was carved out by a Saturnian moon.

Not all asteroids, however, orbit in the main belt between Mars and Jupiter.

▲ When a small body like an asteroid and a big body like a planet come together at regular intervals, the big body repeatedly tugs on the small one with its gravity, eventually kicking the small body from its orbit.

Gravitational dead zones

SHARING THE ORBIT of Jupiter, and perpetually 60° behind or 60° in front of the giant planet as it circles the Sun, are thousands of asteroids. The explanation for these "Trojan asteroids" was provided by the great 18th-century French mathematician Joseph-Louis Lagrange. He realized that, for a small object in a system of two big gravitating bodies, such as the Sun and Jupiter, there are five locations where the combined gravitational forces of the two big bodies provides precisely the force required for the object to rotate with them. The Trojan asteroids occupy two of these Lagrange points, dubbed L4 and L5. A body that becomes trapped in such a gravitational Sargasso Sea can languish there pretty much indefinitely.

Trojan asteroids have been found near other planets, such as Mars and Neptune. Theia, the Mars-mass body thought to have created the Moon when it struck the newborn Earth, is believed to have formed at the L4 or L5 point along the Earth's orbit.

Lagrange points have been exploited by scientific satellites. NASA's Wilkinson Microwave Anisotropy Probe (WMAP) experiment, designed to observe the dim afterglow of the Big Bang, was placed at the L2 point of the Sun–Earth system, 1.5 million kilometers beyond the Earth on the extension of the line between the Sun and Earth. At L2 it could observe the afterglow without being dazzled by the heat of our planet.

In 1975 the L5 Society was founded to promote the ideas of Gerard O'Neill, who fervently believed that one day humans would place a space colony at a Lagrange point.

Trojan asteroids are benign. But another class is potentially deadly.

▼ In the gravitational "force field" of a planet orbiting a star there are five "plateaus," discovered by Joseph-Louis Lagrange, where a third body rotating with the planet can essentially loiter forever. (For the technically-minded, the gridded surface represents the gravitational-inertial potential of the location.)

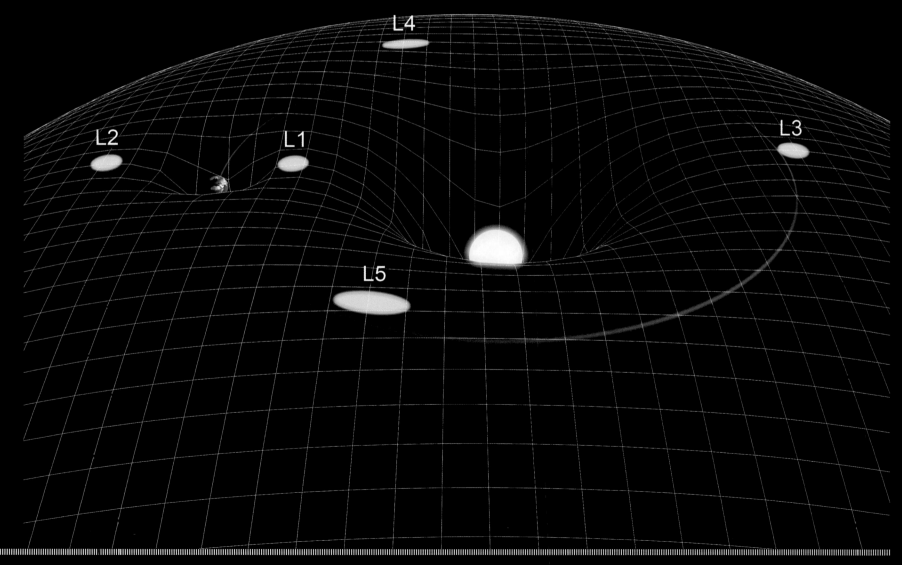

L4

L2   L1   L3

L5

Killer asteroids

OCTOBER 6, 2008, the 60–inch telescope of the Catalina Sky Survey on Arizona's Mount Lemmon picked up an object heading toward the Earth. Nineteen hours later, it plunged through the atmosphere above northern Sudan, the glow of its fireball seen by the Meteosat satellite and the pilot of a KLM flight from Johannesburg to Amsterdam. Fragments of the three-meter rock were later collected.

Object 2008 TC3 is the only object to have been detected in space before it hit the Earth. It is part of a large population of near-Earth objects, or NEOs, whose paths cross the Earth's orbit and could strike the planet. Such bodies have either been knocked out of the main Asteroid Belt by collisions or are rubble from the disintegration of icy comets that ventured too close to the Sun.

Several telescopes have been searching for NEOs since the mid 1990s. The discovery rate, in 2010, was about 700 a year. About 1,000 NEOs of a kilometer or so across are known.

Historically, the Earth has been struck by many such objects. In 1908 a body the size of a terrace of houses exploded high above the Tunguska river in Siberia, flattening 2,000 square kilometers of forest. But 65 million years ago a 10–kilometer body whose impact released millions of times the energy of the biggest H-bomb dealt the dinosaurs a killer blow. The resulting Chicxulub Crater, centered off the coast of Yucatán, is about 180 kilometers across. The impactor would not have been visible until it hit the atmosphere and was heated to incandescence by friction, so the dinosaurs had less than 10 seconds' warning.

▶ T. Rex probably got only 10 seconds' warning of the 10-kilometer asteroid that wiped him out. Until hitting the atmosphere and heating to incandescence, it was invisible.

▲ Asteroids range in size from mere pebbles to minor planets many hundreds of kilometers across. Left to right: Gaspra (18 km long); Eros (33 km); Ida (59 km) and its moon, Dactyl (1.6 km); Mathilde (66 km); and Lutetia (132 km).

Zodiacal dust

IT IS A CLOUDLESS night and you are far from the lights of any city. Look carefully at the sky. If you are lucky, you will notice that in one region the blackness behind the stars is less black than elsewhere. You are seeing the reflection of sunlight from the "zodiacal dust."

The zodiacal dust is a thick and diffuse band of tiny grains swirling around the Sun in the ecliptic, the same plane in which the Earth and planets orbit. Since the planets vacuum up the dust it must be replenished, most probably by debris kicked into space by collisions between rocky bodies within the Asteroid Belt.

The zodiacal dust was the subject of the Ph.D. thesis of rock musician Brian May, who proved its dust particles orbit the Sun in the same direction as the planets. May abandoned his Ph.D. in 1971 to concentrate on his rock group, Queen, but picked it up again and completed his doctorate in 2007.

Fortunately for May, the field of zodiacal dust lay fallow for 36 years. It is now topical because of the search for Earth-like planets around other stars. Such tiny worlds will be super faint compared with their parent stars. However, they will be brighter in a type of light, called far-infrared, given out by cool objects. This is precisely the light with which the zodiacal dust glows. Astronomers are therefore very keen to know whether other stars have zodiacal dust clouds like ours and, if so, what their zodiacal dust clouds will look like from afar. Only then will they be able to pick out planets against the confusing background light.

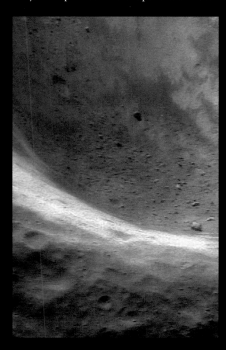

▲ A close-up view of asteroid Eros shows an accumulation of reddish debris including large boulders on the floor of a 5.3-kilometer-diameter crater.

▲ The C–type asteroid Mathilde (left) and the S–type asteroid Eros (right). The color of S-types is consistent with "stony" or rocky composition, whereas C–types are much darker and grayer due to "carbonaceous" materials including carbon compounds.

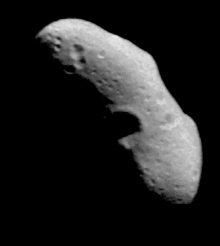

◀ Brian May of Queen. His Ph.D. thesis proved that dust particles orbit the sun in the same direction as the planets.

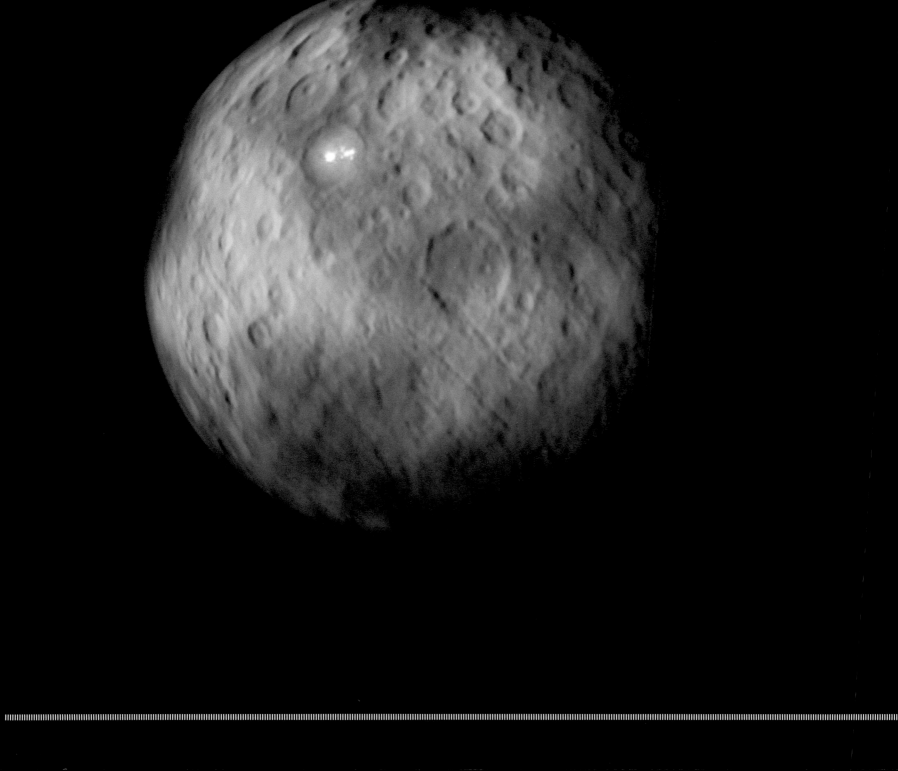

Ceres

CERES IS A spherical ball of rock about the size of the British Isles. For half a century it was hailed as the eighth planet.

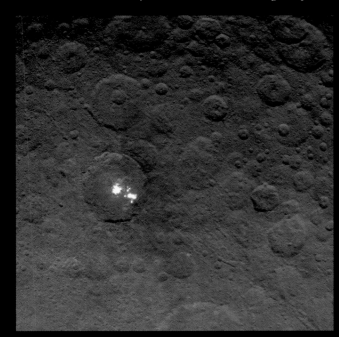

◄ Scientists were puzzled by bright spots found within one of Ceres's craters when NASA's Dawn probe reached the dwarf planet in 2015. Possible explanations include exposures of ice or salt.

ORBITAL DATA

Distance from Sun 381 to 447 million km / 2.55 to 2.99 AU
Orbital Period (Year) 1680.5 Earth days
Length of Day 0.378 Earth days
Orbital Speed 19.4 to 16.5 km/s
Orbital Eccentricity 0.0793
Orbital Incination 10.59°
Axial Tilt 3°

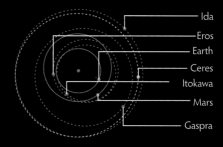

Ida
Eros
Earth
Ceres
Itokawa
Mars
Gaspra

PHYSICAL DATA

Diameter 952 km / 0.075 x Earth
Mass 943,000 million million metric tons
Volume 451 million km³
Gravity 0.028 x Earth
Escape Velocity 0.514 km/s
Surface Temperature 167° to 239°K / −106° to −34°C
Mean Density 2.08 g/cm³

Texas

The planet that never was

IN 1772 Johann Bode, elaborating on the earlier work of Johann Titius and Christian Wolff, presented a curious empirical rule. Take the sequence 0, 3, 6, 12, 24, 48, 96 . . ., where each number is twice its predecessor (except the 3). Add 4 to each, and divide by 10. The resulting sequence (0.4, 0.7, 1.0, 1.6, 2.8, 5.2, 10.0 . . .) turns out to agree pretty well with the distances from the Sun of most planets in "astronomical units" (the distance between Earth and the Sun is 1 AU).

The Titius-Bode Law is a mystery. It may be telling us something important about processes in the protoplanetary disk out of which the planets congealed. But frankly it is not necessary to understand the rule to exploit it.

The rule has an exception. The Titius-Bode Law predicts a planet where none is known–at 2.8 AU from the Sun. In Palermo, Sicily, Giuseppe Piazzi set out to look for such a planet and, on New Year's Day 1801, he found it. Or he thought he had. Ceres, at a mere 950 kilometers across, was a bit on the small side. Worse, its discovery was quickly followed by Pallas (1802); Juno (1804); Vesta (1807); and a host of other asteroids, or "minor planets." Euphoria turned to disappointment.

The irony is that Ceres, once classified as a planet and then downgraded to an asteroid, was, in 2006, upgraded again to a planet. Actually, not quite a planet–an entirely new category of body known as a "dwarf planet."

▼ The two largest asteroids: Ceres (left) and Vesta (right). Ceres is large and spherical enough to qualify as a dwarf planet, but Vesta was shattered by a giant impact billions of years ago.

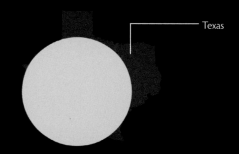

Surface temperature
400°C
200°C
100°C
0°C
800 K
600 K
400 K
200 K
0 K

Mean density
Water 1g/cm³
2g/cm³
Rock 3g/cm³
4g/cm³
5g/cm³
6g/cm³
Iron 7g/cm³

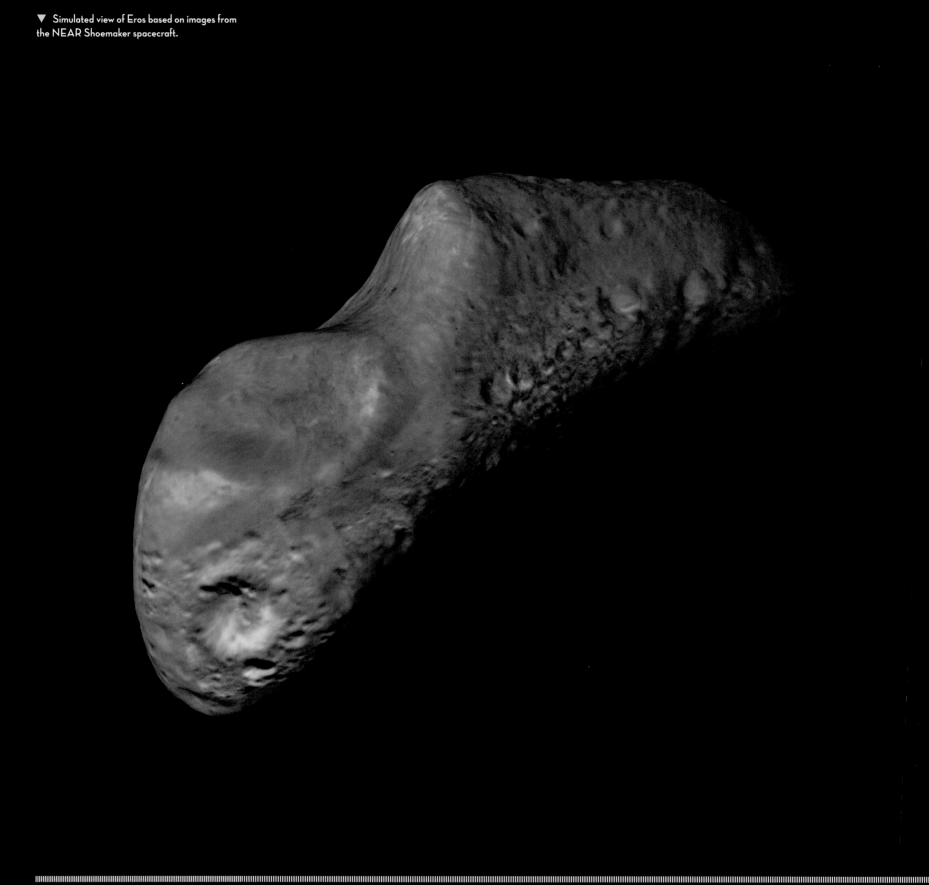

▼ Simulated view of Eros based on images from the NEAR Shoemaker spacecraft.

Eros

EROS, ORBITING BETWEEN the Earth and Mars, has several claims to fame. The first Near-Earth asteroid to be discovered, it could conceivably one day switch to an Earth-crossing orbit. If it ever collided with our planet, the consequences would be dire since Eros is three times bigger than the 10-kilometer asteroid suspected to have ended the age of the dinosaurs. Another of Eros's claims to fame is that it was the first-ever asteroid to gain an artificial satellite. NASA's NEAR-Shoemaker not only orbited the body but, on February 12, 2001, became the first space probe to land on an asteroid.

▲ Asteroid Eros was reached by the NEAR–Shoemaker spacecraft in February 2000, after a four-year journey from Earth.

ORBITAL DATA
Distance from Sun 169 to 267 million km / 1.13 to 1.78
Orbital Period (Year) 643 Earth days
Length of Day 5.27 Earth hours
Orbital Speed 24.36 km/s
Orbital Eccentricity 0.222
Orbital Incination 10.8°

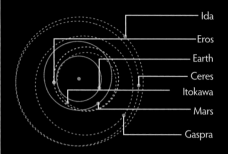

- Ida
- Eros
- Earth
- Ceres
- Itokawa
- Mars
- Gaspra

PHYSICAL DATA
Diameter 16.84 km / 0.001 x Earth
Mass 7 million million metric tons
Volume 2,503 km³
Gravity 0.0006 x Earth
Escape Velocity 0.0103 km/s
Surface Temperature 227°K / -46°C
Mean Density 2.67 g/cm³

Manhattan

▼ The saddle–shaped feature Himeros on the asteroid Eros. Subtle variations in color show older, red–brown debris and the brighter bluer hues of more recent impacts.

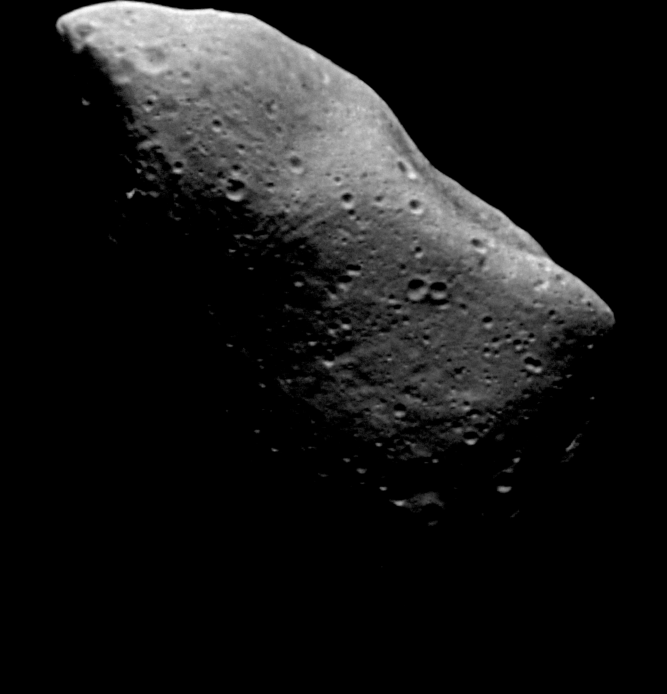

▲ Simulated view of Gaspra based on images
from the Galileo probe, on its way to Jupiter.

Gaspra

GASPRA IS A potato-shaped rock about the size of Manhattan. It was the first asteroid ever approached by a space probe.

Shields up!

ON OCTOBER 29, 1991, NASA's Galileo, en route for Jupiter, flew by Gaspra. On board the space probe was a scientific instrument for measuring magnetic fields. To the scientists' astonishment, the field suddenly leapt in strength at the instant the space probe passed into a region of calm, shielded from the roaring hurricane of the solar wind.

Everyone was amazed because it was believed that only planet-sized bodies could possess magnetospheres, protective magnetic bubbles generated by flows of molten rock that carry electric currents deep in their interiors. Gaspra was too tiny and too solid.

"Mini-magnetospheres" have now been found on the asteroid Ida, on Mars, and on the Moon. On Mars, the localized magnetism may be a memory in the rocks of a planetwide magnetic field that existed long ago in the planet's youth. Similarly, Gaspra and Ida may be fragments of a shattered body that was big enough to have a molten interior and a magnetic field. The Moon's mini–magnetospheres, which have protected the rocks beneath for billions of years, are by contrast believed to have been created by giant impacts.

The exciting thing is that a small astronomical body can possess a protective magnetic shield. Deadly particle radiation from the solar wind has been called the show-stopper for the manned exploration of space-particularly for a six-month flight to Mars. If it proves possible for a spaceship to generate its own mini-magnetosphere, one of the great obstacles to human space exploration may disappear.

ORBITAL DATA

Distance from Sun 273 to 388 million km / 1.82 to 2.59 AU
Orbital Period (Year) 1,199 Earth days
Length of Day 7.042 Earth hours
Orbital Speed 23.9 to 16.8 km/s
Orbital Eccentricity 0.174
Orbital Incination 4.102°
Axial Tilt 72°

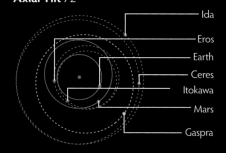

- Ida
- Eros
- Earth
- Ceres
- Itokawa
- Mars
- Gaspra

PHYSICAL DATA

Diameter 12.2 km / 0.001 x Earth
Mass 25 million million metric tons
Gravity 0.004 x Earth
Escape Velocity 0.023 km/s
Surface Temperature 181°K / –92°C
Mean Density 2.70 g/cm³

Manhattan

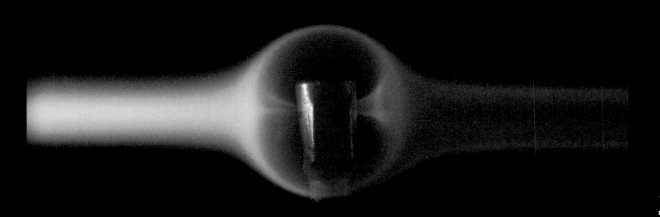

◀ A 2.5-centimeter magnet suspended in a "plasma" wind tunnel shields itself in a protective magnetic bubble.

Surface temperature

400°C — 800 K
200°C — 600 K
100°C — 400 K
0°C — 200 K
0 K

Mean density

Water — 0
1g/cm³
2g/cm³
Rock — 3g/cm³
4g/cm³
5g/cm³
6g/cm³
Iron — 7g/cm³

Ida

IDA WON THE LOTTERY as far as capturing human attention is concerned. Although there are hundreds of thousands of asteroids, only a very few have been visited by space probes. Along with Gaspra, it was imaged by NASA's Galileo space probe en route to Jupiter. The images revealed one of the most heavily cratered bodies in the Solar System, and a big surprise.

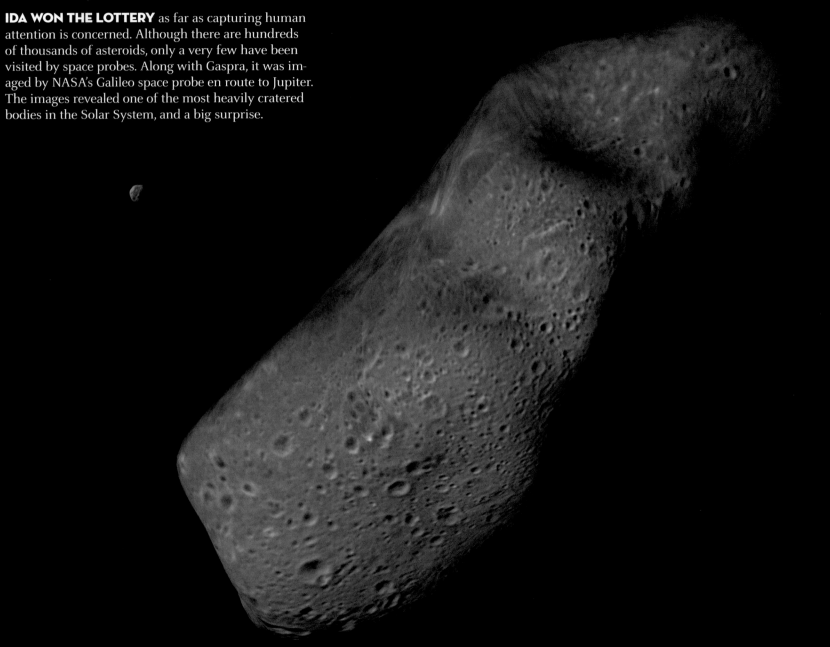

▲ Simulated view of Ida based on images from the Galileo probe, on its way to Jupiter.

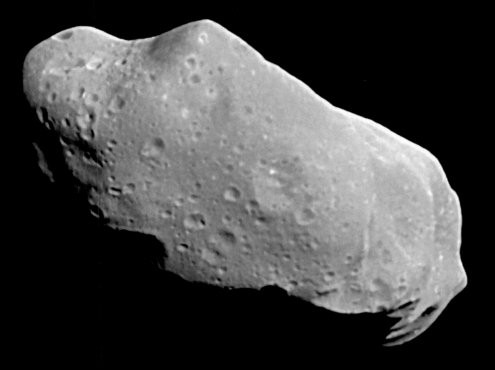

▼ Ida was the first asteroid confirmed to have its own moon. Tiny Dactyl, just 1.4 kilometers across, is visible on the right of this simulated natural color view.

ORBITAL DATA

Distance from Sun 409 to 447 million km / 2.73 to 2.99 AU
Orbital Period (Year) 1768 Earth days
Length of Day 4.63 Earth hours
Orbital Speed 18.4 to 16.8 km/s
Orbital Eccentricity 0.0452°
Orbital Incination 1.14°

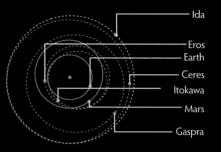

Ida
Eros
Earth
Ceres
Itokawa
Mars
Gaspra

PHYSICAL DATA

Diameter 56 km / 0.004 x Earth
Mass 42 million million metric tons
Volume 16,100 km³
Gravity 0 x Earth
Escape Velocity 0.014 km/s
Surface Temperature 200°K / –73°C
Mean Density 2.60 g/cm³

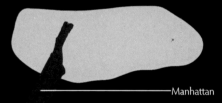

Manhattan

A surfeit of satellites

ONLY BIG THINGS like planets have moons. Right? So imagine the excitement when, on August 28, 1993, NASA's Galileo space probe flew past Ida and took photos that revealed a moon. Barely a twentieth the size of its parent, the 1.4-kilometer Dactyl orbits once every 20 hours at the speed of a slowly thrown baseball. It is not the only moon of a minor body.

Now, almost 200 such moons are known. Several asteroids even boast two. Satellites are surprisingly common. It seems about 2 percent of asteroids have them, and about 10 percent of Kuiper Belt objects. The mystery is how such double and triple systems originated.

One possibility is that the moons are "chips off the old block" formed when an impact broke off a bit of the parent asteroid. There is an alternative: The dark surface of an asteroid absorbs sunlight and heats up more efficiently than lighter parts; that heat, escaping into space, acts like a rocket exhaust, causing the dark areas to recoil more than the light areas. The asteroid spins faster and faster. Since such bodies are thought to be loose rubble piles, held together only by the weak gravity typical of small bodies, a chunk of rubble may spin off and make a moon.

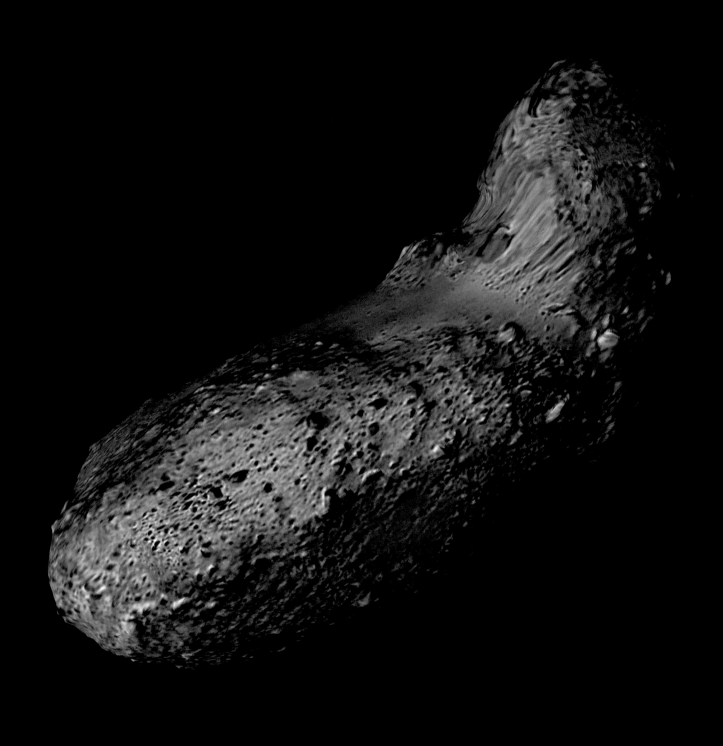

Itokawa

NOTHING LIKE IT has ever been seen before. There are no craters whatsoever on Itokawa, a tiny asteroid the size of a large oil tanker. It appears to be no more than a loose agglomeration of rock and ice chunks. The gravity that holds it together is 100,000 times weaker than on Earth. Like every other asteroid, Itokawa must occasionally be struck by space rocks. Perhaps any crater that forms is filled in when the rubble of the asteroid is jiggled by the gravity of a passing planet–in this case, the Earth.

▼ Itokawa has an extraordinary surface of loosely assembled rubble.

ORBITAL DATA
Distance from Sun 143 to 253 million km / 0.96 to 1.69 AU
Orbital Period (Year) 556 Earth days
Length of Day 12.13 Earth hours
Orbital Speed 34.5 to 19.4 km/s
Orbital Eccentricity 0.28°
Orbital Incination 1.622°

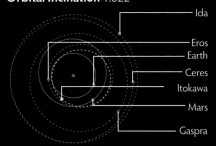

Ida
Eros
Earth
Ceres
Itokawa
Mars
Gaspra

PHYSICAL DATA
Diameter 0.3 km
Mass 35 million million metric tons
Gravity 0.00001 x Earth
Escape Velocity 0.0002 km/s
Surface Temperature 206°K / –67°C
Mean Density 1.90 g/cm^3

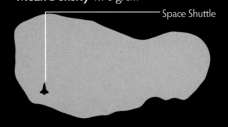

Space Shuttle

Touchdown

THE HYABUSA MISSION is straight from the pages of a science-fiction story. Twice in November 2005 the Japanese spacecraft touched down on the surface of the near-Earth asteroid.

It was a tremendous technological feat. Hyabusa had to race halfway across the Solar System. It had to match its speed with tiny Itokawa, named after Japanese rocket scientist, Hideo Itokawa. It then had to descend to the surface in a maneuver of incredible precision and delicacy.

Hyabusa was not the first spacecraft to land on an asteroid–that honor goes to NASA's NEAR–Shoemaker, which landed on Eros on February 12, 2001. But Hyabusa was the first to take a sample of asteroid dust and return it to Earth. The container came down in the outback of South Australia in the

summer of 2010. Although Hyabusa's dust sampler malfunctioned, scientists still believe it caught dust blown off the surface by its rockets. Such material is important because it will be unchanged since the birth of the Solar System, the kind of stuff from which the Earth itself was formed.

Hyabusa's success is both good and bad news for the human race. One day we may have to deflect an asteroid on a collision course with the Earth. We now know it is possible to land on an asteroid. Unfortunately, the slightest nudge might shatter an asteroid as fragile as Itokawa. Instead of one large space rock heading toward the Earth at 200,000 kilometers an hour, we may succeed only in creating hundreds of smaller rocks heading to the Earth at 200,000 kilometers an hour.

▲ Hideo Itokawa

▲ The Japanese probe Hyabusa soft–landed on Itokawa to take samples of the asteroid's surface back to Earth.

▶ Japanese scientists with the dust sample returned to Earth by the probe.

Surface temperature

800 K
400°C 600 K
200°C 500 K
100°C 400 K
0°C   200 K
       0 K
0
Water 1 g/cm^3
       2 g/cm^3
Rock   3 g/cm^3
       4 g/cm^3
       5 g/cm^3
       6 g/cm^3
Iron   7 g/cm^3

Mean density

The Outer Solar System

NEPTUNE ▶ 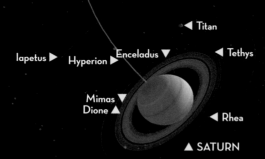 ◀ Triton

Titan ◀

Iapetus ▶   Hyperion ▶   Enceladus ▼   Tethys ◀

Mimas ▼
Dione ▲   Rhea ◀

▲ SATURN

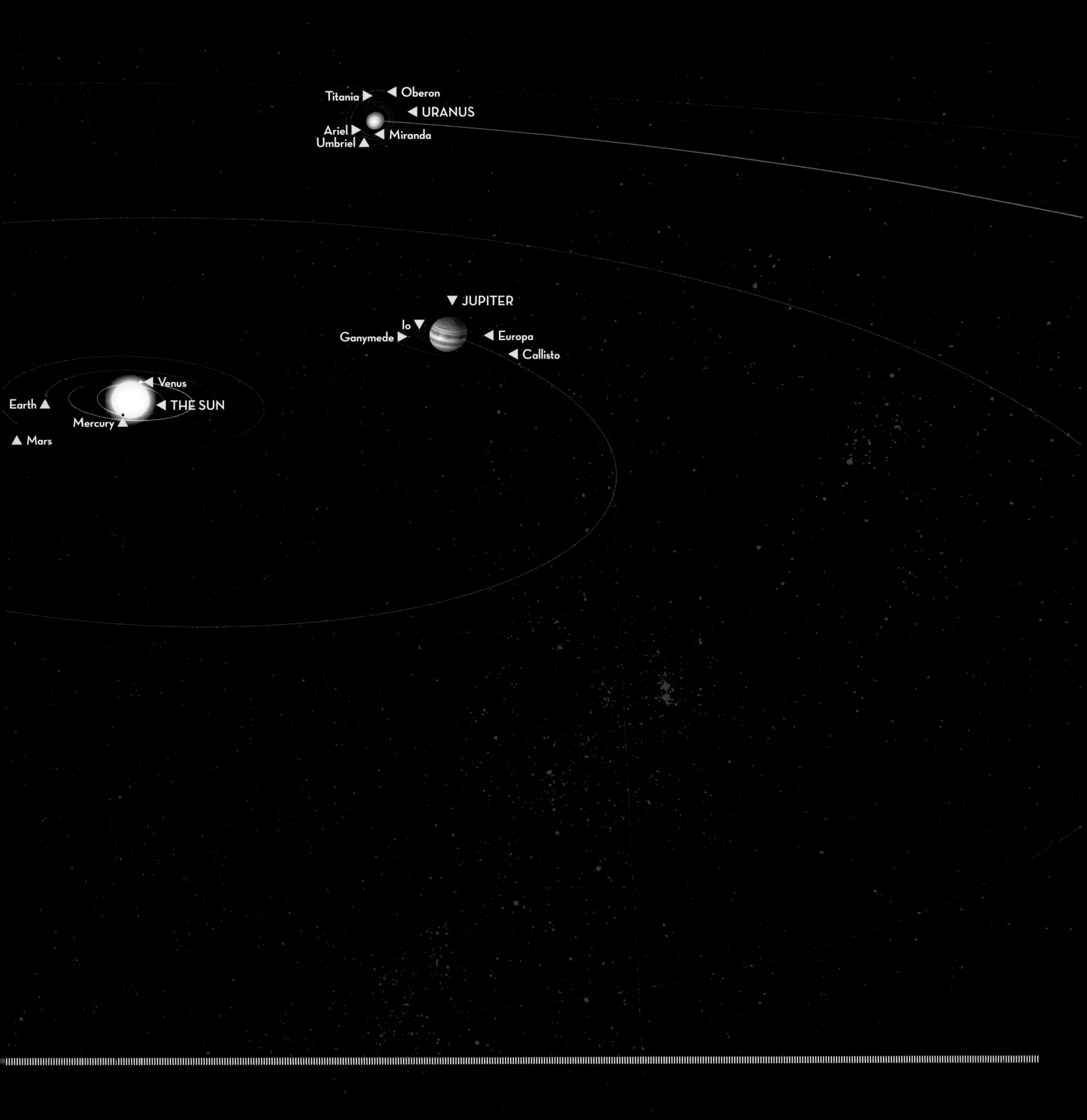

Titania ▶ ◀ Oberon

◀ URANUS

Ariel ▶ ◀ Miranda
Umbriel ▲

▼ JUPITER

Ganymede ▶   Io ▼
                 ◀ Europa
              ◀ Callisto

◀ Venus

Earth ▲   ◀ THE SUN

Mercury ▲

▲ Mars

Jupiter

EVERYTHING ABOUT JUPITER challenges the imagination. It is big enough to engulf 1,300 Earths. It has no surface. It is just a giant ball of gas, held together by gravity and spinning so fast it bulges out at its equator by 7 percent. Staring out of its multicolored, cloud-banded, roiling atmosphere is a malevolent red eye–a hurricane three times bigger than our planet that has not let up in at least 200 years. The planet's entou-rage of moons is nothing short of a mini-Solar System. Worlds in their own right, one is bigger than a planet; another generates more heat, pound for pound, than the Sun; yet another may hide the biggest ocean in the Solar System.

Say you could travel to Jupiter, five times farther from the Sun than the Earth. The first thing you would encounter would be an invisible force field.

ORBITAL DATA
Distance from Sun 741 to 816 million km / 4.95 to 5.45 AU
Orbital Period (Year) 11.86 Earth years
Length of Day 9.93 hours
Orbital Speed 13.7 to 12.4 km/s
Orbital Eccentricity 0.0484
Orbital Incination 1.3°
Axial Tilt 3.12°

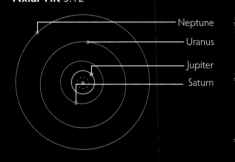
- Neptune
- Uranus
- Jupiter
- Saturn

PHYSICAL DATA
Diameter 142,984 km / 11.2 x Earth
Mass 1,900,000 billion billion metric tons / 318 x Earth
Volume 143,000,000 million km³ / 1,320 x Earth
Gravity 2.39 x Earth
Escape Velocity 59.523 km/s
Surface Temperature 110° to 152°K / –163° to –121°C
Mean Density 1.33 g/cm³

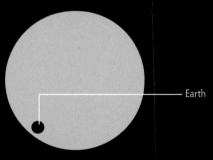
- Earth

ATMOSPHERIC COMPOSITION
Hydrogen 89.8%
Helium 10.2%
Methane 0.3%
Ammonia 0.026%
Hydrogen deuteride 0.003%
Ethane 0.0006%
Water 0.0004%

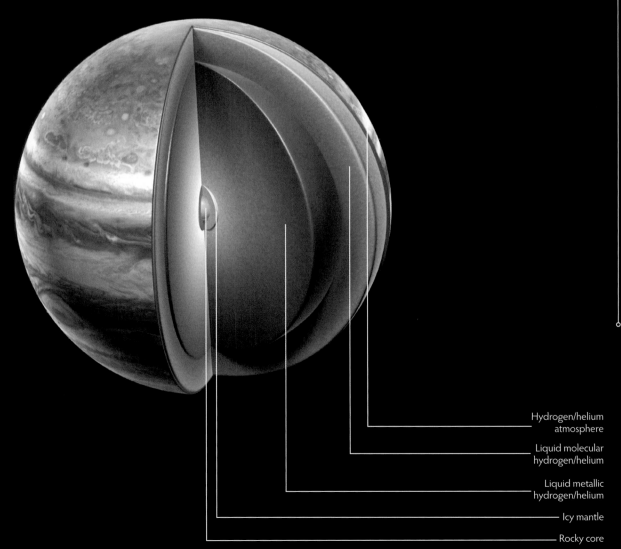

- Hydrogen/helium atmosphere
- Liquid molecular hydrogen/helium
- Liquid metallic hydrogen/helium
- Icy mantle
- Rocky core

Surface temperature: 800 K / 600 K / 400 K / 200 K / 0 K
400°C / 200°C / 100°C / 0°C

Mean density: Iron / 7g/cm³ / 6g/cm³ / 5g/cm³ / 4g/cm³ / 3g/cm³ / Rock / 2g/cm³ / Water / 1g/cm³ / 0

Death Zone

ON NOVEMBER 26, 1995, NASA's Galileo space probe crossed the boundary from interplanetary space into the Jovian magnetosphere, the giant magnetic cocoon around Jupiter. It was still 9 million kilometers from the giant planet–more than 20 times the Earth-Moon distance.

If visible to the naked eye, Jupiter's mammoth magnetosphere would look like a giant elongated teardrop. On the sunward side a spherical cocoon is compressed by the hurricane of particles from the solar wind, and on the opposite side it trails for half a billion kilometers, beyond even the orbit of Saturn.

The space inside the magnetosphere is protected from the particles of the solar wind. But it does contain huge quantities of homegrown, charged subatomic particles. Some are injected by volcanoes on Io, which create a doughnut–shaped ring of sulfur and sodium around Jupiter. The magnetosphere rotates with Jupiter once every 10 hours, raking the inner Galilean moons with deadly particle radiation. The electronics of the Galileo probe had to be specially hardened to withstand the onslaught, which is enough to kill humans quickly.

The poles of Jupiter's magnetic field, which is 15 times stronger than the Earth's, funnel down particles from Io, where they slam into atoms in the planet's atmosphere, creating the most spectacular auroras in the Solar System. Also, as Io cuts through Jupiter's magnetic field, huge electrical currents are generated, as in a dynamo. These surge down through the gas around the moon to Jupiter's atmosphere along magnetic field lines–the path of least resistance–where they create super–bolts of lightning, 100 times as powerful as any on Earth. That atmosphere is something to behold.

▼ A view of the ultraviolet emissions from Jupiter's northern aurora shows the magnetic "footprints" of three of Jupiter's moons. These bright spots and trails are due to charged particles from the moons flowing in an electric current along the field lines of Jupiter's magnetosphere.

▲ Ultraviolet light shows details of Jupiter's auroral emissions at both poles. Curtains of light stretch hundreds of kilometers above the edge of the planet. A distinct spot and trail marks the point where charged particles from Jupiter's moon Io enter the Jovian atmosphere.

▲ Jupiter displays polar aurorae a thousand times more powerful than those on Earth—so powerful that they produce X-ray emissions, shown here using data from the Chandra X-ray Observatory satellite.

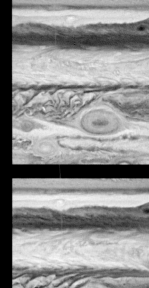

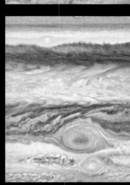

▶ In recent years, the Hubble Space Telescope has allowed more frequent views of Jupiter's atmospheric features. This sequence of images, taken from May to July, 2008, shows the Great Red Spot devouring a smaller red storm system approaching from the west.

◀ Jupiter's Great Red Spot has been observed for at least 150 years. At that time it was twice as large as it is today.

Eye of the tiger

IMAGINE A STORM that has been raging not for days but for centuries. Jupiter's Great Red Spot, possibly seen as early as 1655, stupefies the imagination. It could swallow 100,000 Hurricane Katrinas.

The Red Spot spins once every six days and towers eight kilometers above the surrounding cloud tops. Astronomers once thought it was a vortex of gas swirling around a titanic mountain, but Jupiter is a liquid-gas planet with no solid surface. The Red Spot is actually the top of a rising mass of gas about which the atmosphere circulates. The Red Spot's color is thought to come from chemicals like phosphorus dredged from below.

When another Jovian storm forms, it is eaten by the Red Spot. The Spot's stability probably indicates it is a "soliton." Such a "solitary wave," able to regenerate itself as fast as it

dissipates, was first observed by scientist John Scott Russell on the Union Canal near Edinburgh in 1834. When a boat stopped, the bow wave kept going, and Russell followed it for two miles on horseback.

Jupiter sports many parallel bands of cloud. The bright "zones" are rising gas, and the dark "belts" descending gas. Jupiter's atmosphere is being heated from below by its own warmth, so we may be seeing the equivalent of convection in water being heated in a saucepan. The banded pattern is created by the fast spin of the Jovian atmosphere, 30 times faster at the equator than the Earth's. Similar banded patterns have been observed in the laboratory in the convection of rapidly rotating fluids.

So much for Jupiter's atmosphere, but what is it like inside the planet?

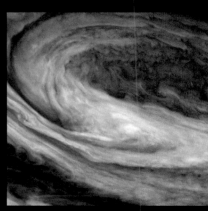

▲ The Great Red Spot is an enormous long-lived storm, three times the size of the Earth, rotating anti-clockwise once every six days.

Journey to the center of Jupiter

ON SEPTEMBER 21, 2003, NASA's Galileo space probe, traveling at 30 miles a second, crashed into the far side of Jupiter. It barely penetrated the super–dense Jovian atmosphere before flashing into a fireball. But what if you could set the controls for the heart of Jupiter? What might you find if you are impervious to the violent updrafts and swirling super-hurricane-force winds?

As you plunge through the cloud banks, sunlight dwindles to a distant memory and only stuttering lightning provides illumination. Some scientists have suggested that Jupiter's clouds might be home to jellyfish-like creatures, drifting on the winds, suspended beneath balloon-like bags of gas. But that is truly speculative.

Jupiter's atmosphere has no solid base like the Earth's. Nevertheless, after about 1,000 kilometers the weight of the atmosphere crushing down from above begins to turn the hydrogen gas to liquid. You plunge through a liquid denser than any solid on Earth. Not for thousands of kilometers but for *tens of thousands* of kilometers. Eventually, the crush is so great it turns the hydrogen into something without precedent: liquid metallic hydrogen. Gigantic electric currents circulating in this conducting medium, like currents circulating in the Earth's molten interior, are believed to generate Jupiter's tremendous magnetic field.

Still deeper, the temperature rises to 30,000°C, and ahead of you at last is solid rock. Compressed to only about one and a half times the size of the Earth yet containing about 20 times its mass, this is the rocky "seed" that gathered about it a mantle of gas to form Jupiter. The journey is over. You have reached Jupiter's heart of darkness.

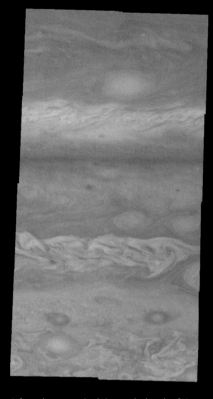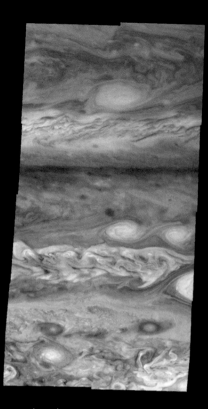

▲ Infrared imagery (right) reveals details of Jupiter's atmosphere that are not seen at visible wavelengths (left). High, thick clouds appear white, high thin clouds are light blue, deep clouds are red, and deep, clear areas are overlaid by a purple haze.

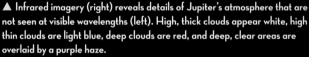

◀ This infrared image highlights variations in Jupiter's upper atmosphere. It is a remarkably sharp view for a ground-based telescope. Using a technique called "adaptive optics," the telescope's mirror is distorted during the exposure to compensate for atmospheric disturbances.

▲ Icy fragments of Shoemaker-Levy-9 plunge into Jupiter's atmosphere.

▲ Impact scars from Shoemaker–Levy 9 fragments G and L, taken on the July 21, 1994.

Protector or enemy?

IN 1994 COMET Shoemaker-Levy shattered. Jupiter's powerful gravity tore it into fragments, stringing them out like a line of pearls. Astronomers had a ringside seat as the pieces headed for the giant planet, where each slammed into the Jovian atmosphere with the energy of hundreds of thousands of one-megaton H-bombs. Never before had such a series of impacts on another planet been witnessed.

In 1686 the Italian astronomer Giovanni Cassini drew an event that looked like an impact. If he really did see one, then it was the only such observation for around two centuries. Techniques have now improved so greatly that three such impacts were spotted between July 2009 and August 2010.

It seems that Jupiter's powerful gravity can not only break up a comet

that ventures in from the cold beyond the planets but suck in the pieces too. In doing so, the "biggest target in the Solar System" mops up bodies that could potentially strike the Earth. Jupiter is our protector . . . Or is it?

Initially, debris in the outer Solar System was safely confined to circular orbits around the Sun. However, over time the gravitational tug of Jupiter changed these paths, dragging material

into the inner Solar System.

So Jupiter is both good and bad. It mops up some hazardous bodies but sends others our way with the potential to trigger mass extinctions. It is the gatekeeper between the inner and outer Solar System.

It is sobering to think that a body about 400 million miles away has such a profound influence on life on Earth. And not just the Earth.

Lord of chaos

JOHANNES KEPLER discovered that the orbits of the planets around the Sun are ellipses, and Isaac Newton showed that this was exactly what was predicted by his theory of gravity. But planets do not move in ellipses–not exactly.

Newton knew his result was an approximation. Gravity is not a force between a big thing like the Sun and a small thing like a planet; it is a universal force between *all masses*. There is a force of gravity between you and the coins in your pocket, between you and your office mate, and between any planet and every other planet.

The tug on a planet by other planets perturbs its orbit from a perfect ellipse. And Jupiter, being the most massive planet, has the biggest reach. Its tug even makes the Sun wobble. Similar wobbles of other stars signal the presence of unseen planets around them.

Over time, these tiny perturbations can build up. Like the flap of a butterfly's wings in Fiji eventually causing a hurricane in the Caribbean, they can spawn something unpredictable and dramatic. The effect is known as "dynamical chaos." Even though the Solar System appears to run by Newtonian clockwork, at some time in the future the clockwork may suddenly go haywire, perhaps even ejecting a planet like the Earth into interstellar space.

In fact, something like this happened 3.8 billion years ago. Jupiter and Saturn entered a resonance in which their orbital periods were in the ratio 1:2. The effect was to turn the inner Solar System into a shooting gallery. The giant Mare basins on the Moon are scars from this Late Heavy Bombardment.

▲ If Jupiter had been bigger, the Earth may have had two suns.

Failed star

IN ARTHUR C. CLARKE'S novel *2010: Odyssey Two*, alien machines turn Jupiter into a star. Far-fetched, you may think. But Jupiter *is* a giant ball of gas like the Sun. Could it have become a star?

The central characteristic of a star is that it generates its own light and heat. The central characteristic of a planet is that it is warmed by a star and shines only by reflected light. What then of Jupiter? Measurements show it radiates into space about twice as much heat as it receives from the Sun.

Astronomers believe the interior of Jupiter is slowly contracting–at about a millimeter a year–and this is converting gravitational energy into heat energy. This is not what the Sun is doing. Its heat is a by-product of nuclear reactions–specifically, the "fusion" of hydrogen nuclei into helium nuclei. But these are triggered only at a temperature of at least 10 million degrees Celsius. Jupiter simply has insufficient matter crushing down on its core to squeeze it to such a dizzyingly high temperature.

How much mass is needed to become a star? The answer is about 80 times the mass of Jupiter, or about 8 percent of the mass of the Sun. Jupiter therefore did not make the grade as a second Sun.

Astronomers recognize planets and stars and a gray–or, rather, brown–area in between. Bodies ranging in mass from about one to 80 Jupiters are "brown dwarfs," or failed stars. Strictly speaking, Jupiter is not a planet at all but a brown dwarf.

▲ Scars from the Shoemaker–Levy 9 impacts evolved in the days and weeks following the comet's collision with Jupiter. This image shows eight impact sites (from left): E/F, H, N, Q1, Q2, R and D/G.

▲ The Voyager probes discovered a faint ring system when they visited Jupiter in 1979. This image was taken looking back at the limb of Jupiter from within its shadow.

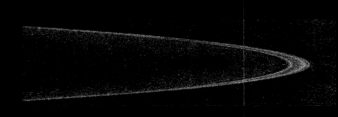

▲ The clearest pictures of Jupiter's ring system were obtained in 2007 by the New Horizons probe, passing Jupiter on its way to Pluto and the Kuiper Belt. This image records back–scattered light as the probe approached Jupiter from the direction of the Sun.

▲ Jupiter map based on images from the Cassini mission. (Mollweide projection map centered on 0° longitude.)

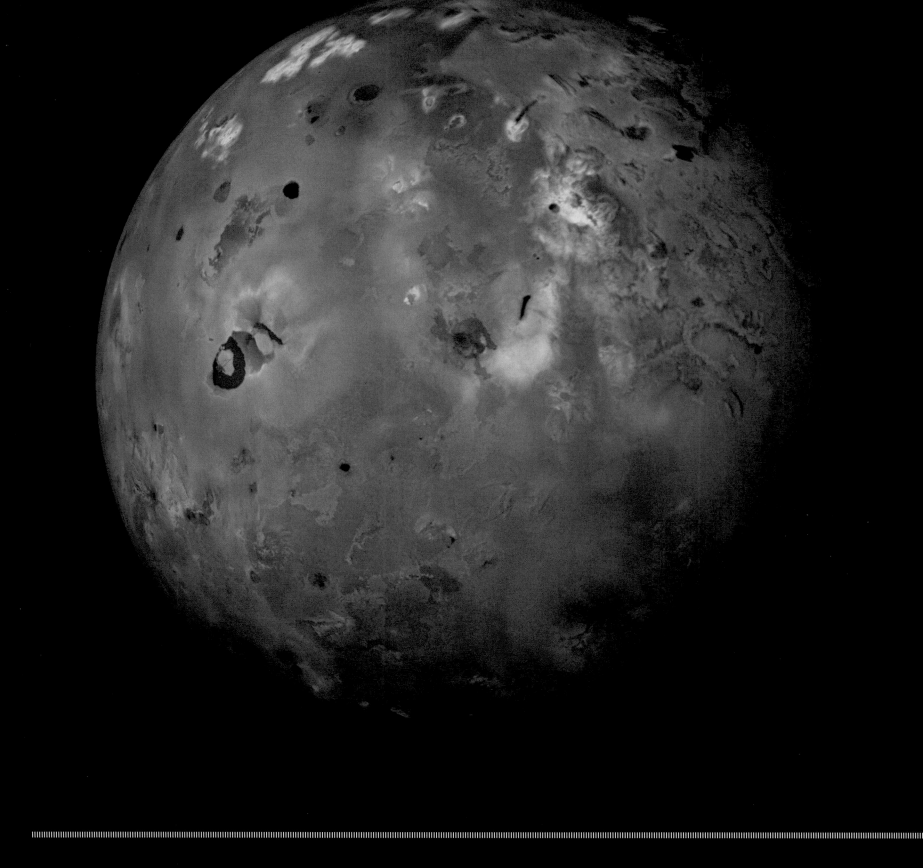

Io

IMAGINE A MOON pretty much the same size and composition as our own Moon but whose surface looks like a melted pizza. Imagine a moon that orbits as far away from its planet as the Moon is from the Earth, yet whirls around that planet not in 28 days like the Earth's satellite but in a mere 1.7 days. Welcome to Io, the innermost of Jupiter's four giant Galilean moons.

Like our own Moon, Io is tidally locked, one face forever toward its parent. If you could stand on that face and look up, what a view you would have. Jupiter and its whirling, multicolored cloud belts would fill almost a quarter of the sky. In the life of this moon, Jupiter is everything. It dominates Io as no other planet dominates a satellite. It is a domination that has had extraordinary and dramatic consequences for the pizza moon.

▼ Io map based on images from the Galileo and Voyager missions. (Mollweide projection map centered on 0° longitude.)

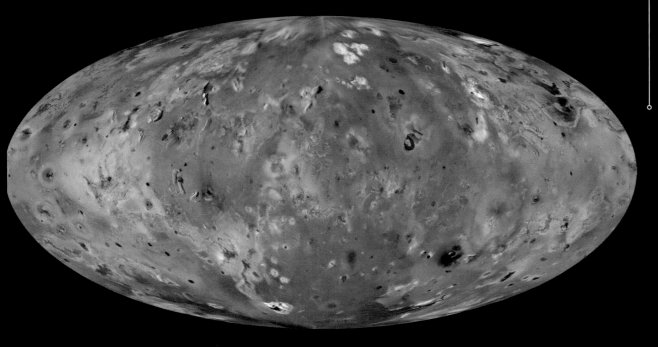

ORBITAL DATA
Distance from Jupiter 420,000 to 423,000 km
Orbital Period (Year) 1.77 Earth days
Length of Day 1.77 Earth days
Orbital Speed 17.4 to 17.3 km/s
Orbital Eccentricity 0.004
Orbital Incination 0.05°
Axial Tilt 0°

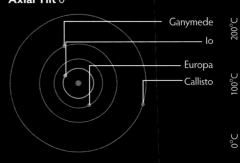

Ganymede
Io
Europa
Callisto

PHYSICAL DATA
Diameter 3,643 km / 0.28 x Earth
Mass 89 billion billion metric tons / 0.01 x Earth
Volume 25,300 million km³ / 0.02 x Earth
Gravity 0.183 x Earth
Escape Velocity 2.558 km/s
Surface Temperature 90° to 130°K / −183° to −143°C
Mean Density 3.57 g/cm³

The Moon

ATMOSPHERIC COMPOSITION
Sulphur dioxide 90%
Sulphur monoxide 3%
Sodium chloride 3%
Sulphur 2%
Oxygen 2%

The hottest pizza

IT IS MARCH 8, 1979. Voyager 1, flying faster than a speeding bullet, has streaked through the Jupiter system and is now heading toward its Fall 1980 rendezvous with Saturn. Before the probe leaves the giant planet behind forever, its camera points backward and takes a parting shot–of Io. When navigation engineer Linda Morabito studies the image, her heart misses a beat. Spouting from the tiny crescent moon, silhouetted against the starry backdrop of space, is a phosphorescent plume of gas.

Morabito has discovered the super-volcanoes of Io. Over the next days, the Voyager team, poring over image-enhanced photos and thermal data, will discover a total of eight gigantic plumes, pumping out matter hundreds of kilometers into space.

Io, it turns out, is the most geologically active body in the Solar System, with more than 400 active volcanoes. Its orange and yellow and brown "pizza" surface is peppered with vents, reminiscent of the geysers of Yellowstone Park.

Actually that is what they are: geysers rather than volcanoes. Lava from the moon's molten interior, instead of erupting directly, superheats liquid sulfur dioxide just beneath the surface, which bursts from the vents like the pressurized steam of a terrestrial geyser.

Io is pumping about 10,000 million tons of matter into space every year. As the matter falls back in the moon's low gravity, it coats the surface with sulfur, just like the deposits around a Yellowstone fumarole. The colors of the pizza moon are simply the phases sulfur exhibits at different temperatures.

But what is powering Io's super-volcanoes?

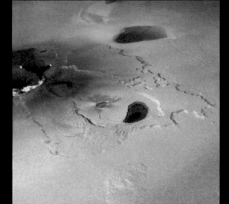

◄ A gash in Io's crust reveals the glow of hot molten lava as it erupts from an active volcano. The surrounding plains and plateaus are silicate rock coated with yellow sulphur, peppered with dark pits of volcanic material.

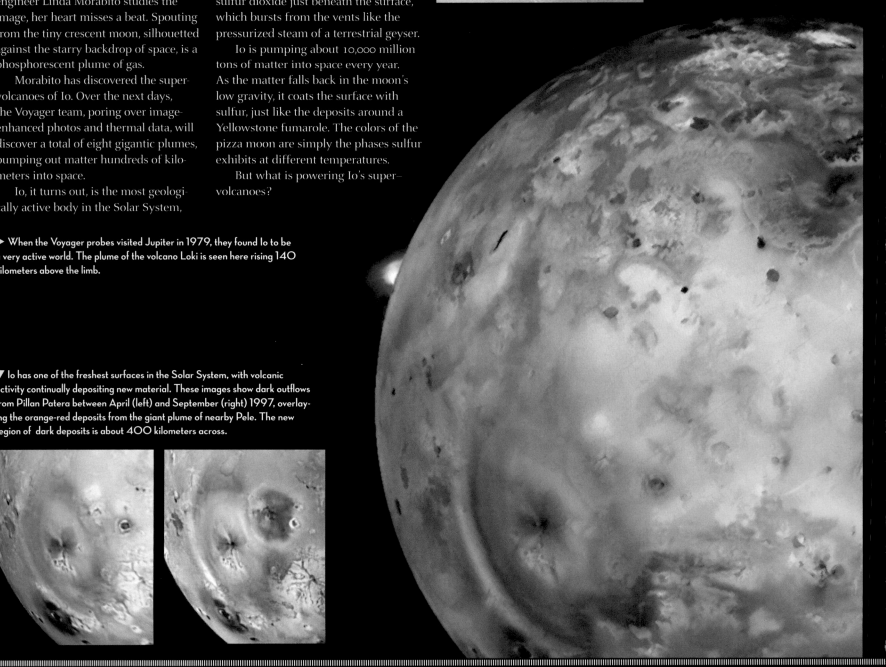

► When the Voyager probes visited Jupiter in 1979, they found Io to be a very active world. The plume of the volcano Loki is seen here rising 140 kilometers above the limb.

▼ Io has one of the freshest surfaces in the Solar System, with volcanic activity continually depositing new material. These images show dark outflows from Pillan Patera between April (left) and September (right) 1997, overlaying the orange-red deposits from the giant plume of nearby Pele. The new region of dark deposits is about 400 kilometers across.

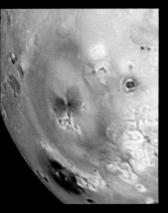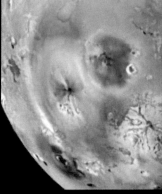

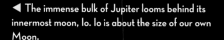
◄ The immense bulk of Jupiter looms behind its innermost moon, Io. Io is about the size of our own Moon.

Please squeeze me

IF YOU SQUEEZE a rubber ball repeatedly, its interior gets hot. This is what Jupiter's gravity is doing to its innermost giant moon, Io. The gas giant's pull on the near side of the moon is stronger than on the far side, so the moon is stretched. And, just like a piece of toffee, if it is stretched in one direction, this is compensated for by compression in the perpendicular direction.

Io orbits as far from Jupiter as the Moon does from Earth. But Jupiter is about 2,600 times as massive as the Moon. Consequently, the tidal stretching and squeezing of Io by Jupiter is about 2,600 times bigger than the tidal stretching of the Earth by the Moon. If the Earth were in Io's place, its tides, rather than changing the sea level by meters, would be kilometers high.

In fact it is worse than this. For every four orbits of Jupiter made by Io, Europa completes two and Ganymede one. The existence of this orbital resonance means that, periodically, the satellites align themselves in such a way that Io is pulled one way by the moons and the opposite way by Jupiter. With such a tug-of-war going on, no wonder Io's interior, volume for volume, contains more heat than the Sun.

Before Voyager, people thought that because Io is close in size to the Moon, they were going to see a world like the Moon. Except Stanton Peale: a week before Voyager 1's encounter with Jupiter, the California-based physicist published a paper suggesting that tidal heating might create volcanism on Io.

◄ Io casts a shadow across the face of Jupiter in this image taken by the Cassini probe, on its way to Saturn in 2000.

Europa

EUROPA IS SMOOTH. Really smooth. From a distance, it looks exactly like a pool cue ball. Say you took a pen and scribbled on the surface of such a ball. If Europa were shrunk to cue-ball size, not a single feature on its surface would be as high as the thickness of the ink. Devoid of mountains or valleys–or even craters–Europa is the biggest ice rink in the Solar System.

Jupiter's ice moon, racing around Jupiter once every 3.6 days, is the second most distant of its four giant Galilean moons. Although it looks featureless and dull from afar, close up it is a very different world indeed. In fact, after Mars, Europa is considered the most exciting world in the Solar System.

▼ Europa map based on images from the Galileo and Voyager missions. (Mollweide projection map centered on longitude 90° East.)

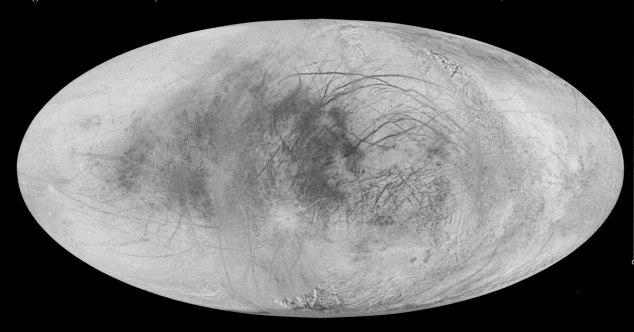

ORBITAL DATA
Distance from Jupiter 664,000 to 678,000 km
Orbital Period (Year) 3.55 Earth days
Length of Day 3.55 Earth days
Orbital Speed 13.9 to 13.6 km/s
Orbital Eccentricity 0.0101
Orbital Incination 0.47°
Axial Tilt 0.1°

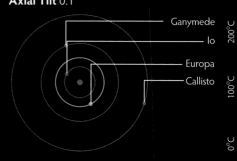

Ganymede
Io
Europa
Callisto

PHYSICAL DATA
Diameter 3,122 km / 0.25 x Earth
Mass 48 billion billion metric tons / 0.01 x Earth
Volume 15,900 million km³ / 0.01 x Earth
Gravity 0.134 x Earth
Escape Velocity 2.026 km/s
Surface Temperature 50° to 125°K / -223° to -148°C
Mean Density 3.02 g/cm³

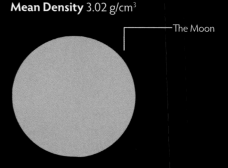

The Moon

ATMOSPHERIC COMPOSITION
Oxygen 100%

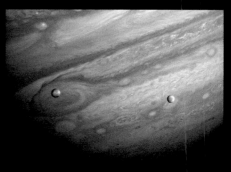

▶ Jupiter's bright, bland moon Europa contrasts with its more colorful sister, Io in this image taken by Voyager 1 in February 1979.

Ice or ocean?

IN 1979, to everyone's surprise, NASA's Voyager 2 space probe revealed that Europa's icy surface was in fact covered in a vast and complex network of cracks and ridges. The big question was: why?

Later, more detailed images taken by the Galileo probe again showed a fragmented surface. It bore a striking resemblance to sea ice in the Arctic–ice that has shattered into jagged pieces and drifted on currents before being glued back together again. For many planetary scientists, this similarity is too much of a coincidence. If similar processes are operating on Europa, it implies the existence of an ocean under the surface.

There is only one other place in the Solar System known to have oceans: Earth. Imagine, then, the jubilation of planetary scientists at the prospect of water on Jupiter's moon. But does Europa really have a subsurface ocean?

Europa orbits only slightly further from Jupiter than Io. The tidal stretching forces of the giant planet, which have turned Io's rocky innards molten, should have melted Europa's icy interior. Observations from NASA's Galileo probe indicate that Europa's interior and crust rotate at different speeds. This is further evidence that the crust may be floating on liquid.

Europa shows every sign that kilometers beneath its icy crust is an ocean, itself perhaps as much as 100 kilometers deep. That would make it the biggest ocean in the Solar System. An ocean raises the tantalizing possibility of life.

▶ Europa has a thick icy crust overlying a rocky interior, possibly with a liquid ocean between the two. Great linear fractures are thought to mark where material from the interior has welled up to the surface.

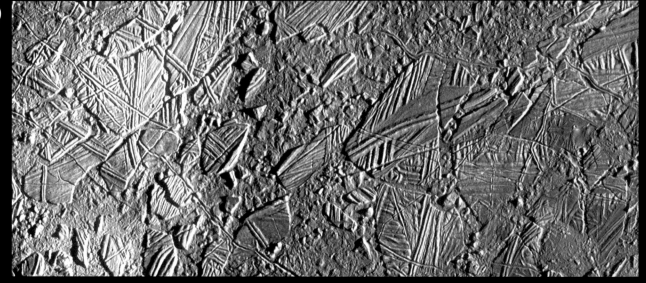

▲ Conamara Chaos shows fractured, displaced, and tilted rafts of ice, not unlike areas of sea ice back on Earth–strong evidence that the ice crust may be underlain by a liquid ocean. Fine-grained ice appears white, coarser-grained ice appears blue, and non–ice material appears red-brown.

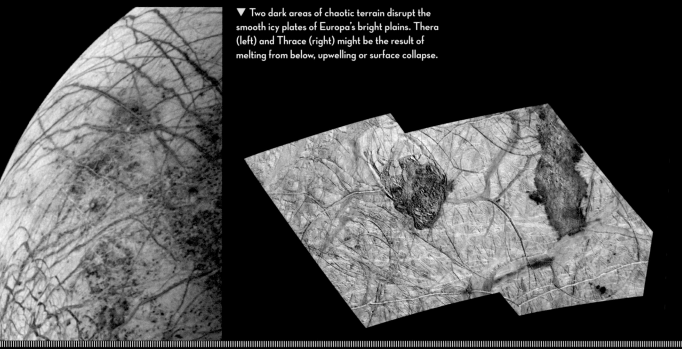

▼ Two dark areas of chaotic terrain disrupt the smooth icy plates of Europa's bright plains. Thera (left) and Thrace (right) might be the result of melting from below, upwelling or surface collapse.

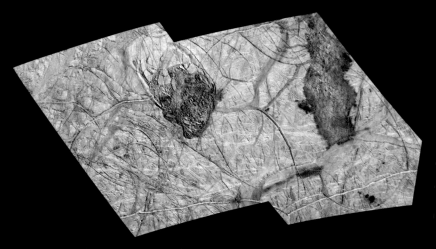

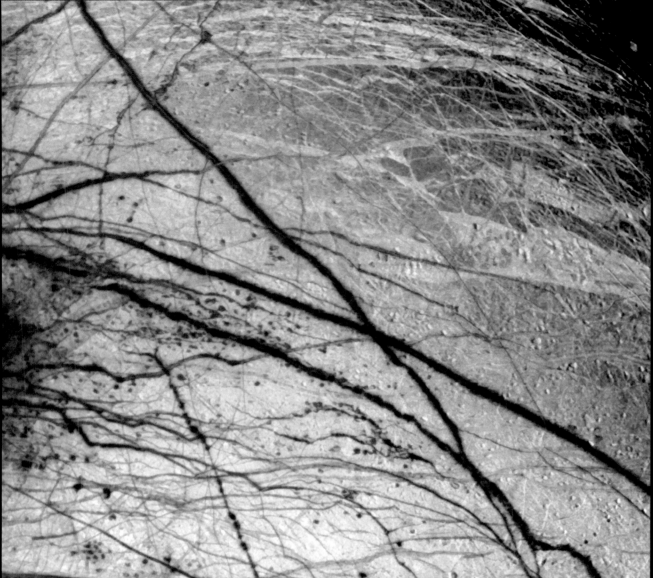

◀ Europa's surface can be divided into three terrains: bright plains, appearing white and blue in this color-enhanced image; darker mottled terrain, appearing red-brown; and darker linear fractures (also red-brown).

▼ The thriving ecosystems around these volcanic events on the seafloor give biologists hope that life may survive in inhospitable places in the Solar System.

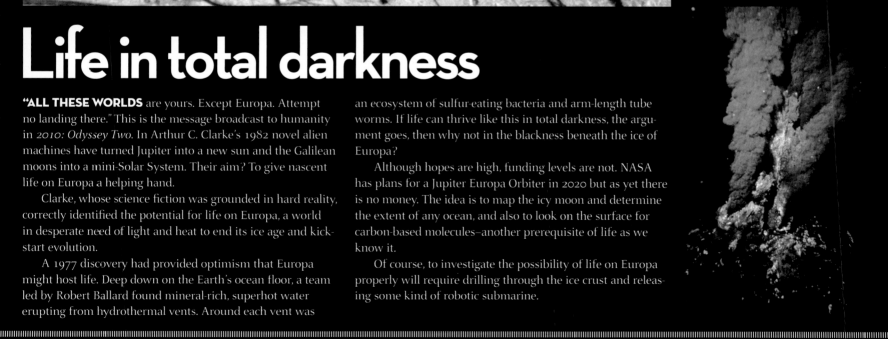

Life in total darkness

"ALL THESE WORLDS are yours. Except Europa. Attempt no landing there." This is the message broadcast to humanity in *2010: Odyssey Two*. In Arthur C. Clarke's 1982 novel alien machines have turned Jupiter into a new sun and the Galilean moons into a mini-Solar System. Their aim? To give nascent life on Europa a helping hand.

Clarke, whose science fiction was grounded in hard reality, correctly identified the potential for life on Europa, a world in desperate need of light and heat to end its ice age and kick-start evolution.

A 1977 discovery had provided optimism that Europa might host life. Deep down on the Earth's ocean floor, a team led by Robert Ballard found mineral-rich, superhot water erupting from hydrothermal vents. Around each vent was

an ecosystem of sulfur-eating bacteria and arm-length tube worms. If life can thrive like this in total darkness, the argument goes, then why not in the blackness beneath the ice of Europa?

Although hopes are high, funding levels are not. NASA has plans for a Jupiter Europa Orbiter in 2020 but as yet there is no money. The idea is to map the icy moon and determine the extent of any ocean, and also to look on the surface for carbon-based molecules–another prerequisite of life as we know it.

Of course, to investigate the possibility of life on Europa properly will require drilling through the ice crust and releasing some kind of robotic submarine.

Ganymede

MOONS ARE SMALLER than planets. Right? Not in the case of Jupiter's Ganymede, which is bigger than Mercury. Measurements made by NASA's Galileo space probe indicate that the giant moon has a metal and rock interior encased in a thick icy crust. That crust is heavily cratered and crossed by strange grooves and ridges, which have been described as like tire tracks in the desert. These suggest that the crust is shearing and buckling and twisting, something that could happen if, as on Earth, the material beneath is flowing like a liquid. That liquid may be slushy ice.

Ganymede whirls around Jupiter in only seven days. It is the third farthest out of the four moons discovered by Galileo in January 1610. Remarkably there may once have been many more of these giant satellites.

▼ Ganymede map based on images from the Galileo and Voyager missions. (Mollweide projection map centered on longitude 180° West.)

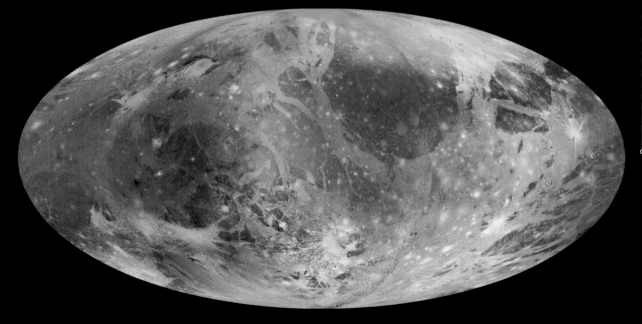

ORBITAL DATA
Distance from Jupiter 845,000 to 1,290,000 km
Orbital Period (Year) 7.15 Earth days
Length of Day 7.15 Earth days
Orbital Speed 13.5 to 8.8 km/s
Orbital Eccentricity 0.21
Orbital Incination 0.2°
Axial Tilt 0.33°

Ganymede
Io
Europa
Callisto

PHYSICAL DATA
Diameter 5,262 km / 0.41 x Earth
Mass 148 billion billion metric tons / 0.02 x Earth
Volume 76,300 km³ / 0.07 x Earth
Gravity 0.146 x Earth
Escape Velocity 2.742 km/s
Surface Temperature 70° to 152°K / −203° to −121°C
Mean Density 3.02 g/cm³

The Moon

ATMOSPHERIC COMPOSITION
Oxygen 99.999%
Hydrogen 0.001%

◀ Jupiter is about to eclipse Ganymede in this view from the Hubble Space Telescope. The image was taken to take advantage of the light reflected from the moon to examine haze layers in Jupiter's upper atmosphere.

Big planet ate my moon

IN THE EARLY days of the Solar System, Jupiter cannibalized 20 or more of its moons. Today's Galilean satellites are the last surviving generation.

This conclusion comes from simulations by Robin Canup and William Ward of the Southwest Research Institute in Boulder, Colorado. Although Jupiter and its moons look like a "mini-Solar System," they say, there is a difference. Whereas the Sun and its surrounding debris disk formed first, with the planets congealing later, the disk around Jupiter formed essentially at the same time as the moons. Consequently, the growing moons interacted with the disk material still being sucked in from the rest of the Solar System.

The gravity of the growing moons stirred up the disk, causing "spiral density waves" to ripple through it. Interactions between the moons and the spiral waves pushed the

moons toward Jupiter. The bigger the moon, the bigger the effect. So, when a moon reached a critical mass, it was swallowed.

Once one set of moons was swallowed, say Canup and Ward, a new set started forming. There may have been five generations. The current Galilean moons survived because as they formed, the inflow of material into the disk from the Solar System choked off.

The total mass of the moons in each generation was the same, but the number of moons could have varied. At one time Jupiter may have had five moons; at another, just one. Something similar may have happened around Saturn. There, the last generation contained just one moon: Titan.

Jupiter's Galilean moons have played a key role in science: they allowed an estimation of the speed of light.

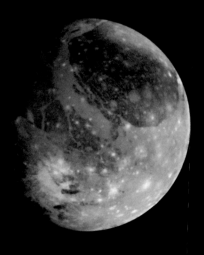

▲ A closer look at Ganymede shows the dark regions to be more heavily cratered, therefore older, than the brighter ridged and grooved regions.

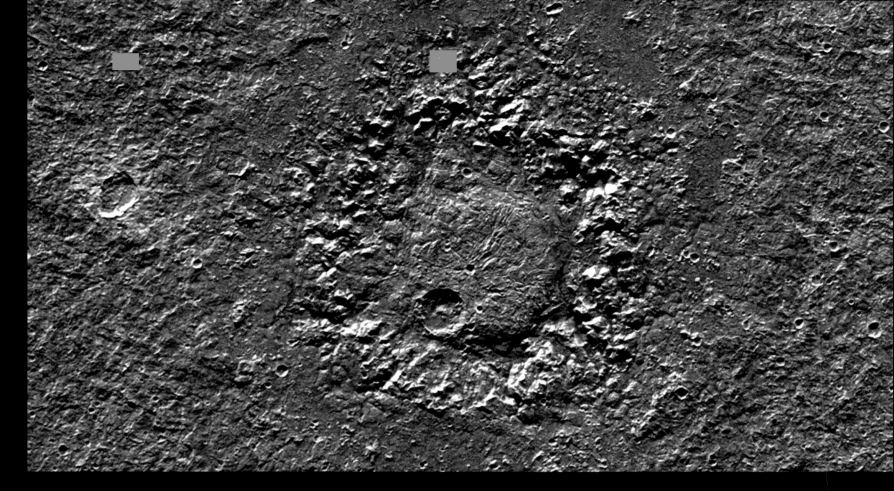

The speed of light

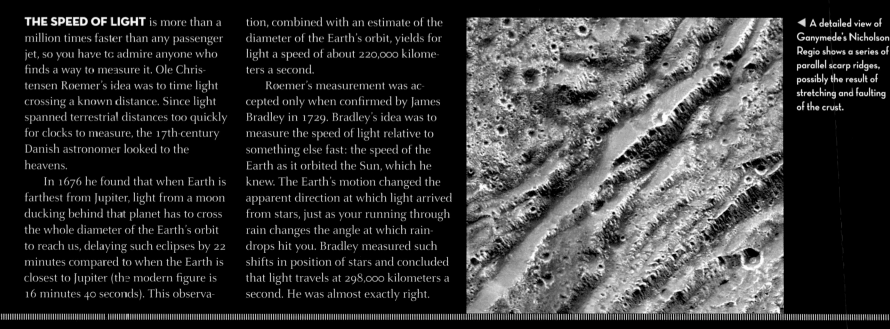

▲ Ganymede's crater Neith has an unusual shape, with a large central dome, a rugged ring, and a much more subdued outer rim. The shape could be due to a high-energy impact in a relatively weak surface, or to long-term post-impact relaxation of a viscous material.

THE SPEED OF LIGHT is more than a million times faster than any passenger jet, so you have to admire anyone who finds a way to measure it. Ole Christensen Røemer's idea was to time light crossing a known distance. Since light spanned terrestrial distances too quickly for clocks to measure, the 17th-century Danish astronomer looked to the heavens.

In 1676 he found that when Earth is farthest from Jupiter, light from a moon ducking behind that planet has to cross the whole diameter of the Earth's orbit to reach us, delaying such eclipses by 22 minutes compared to when the Earth is closest to Jupiter (the modern figure is 16 minutes 40 seconds). This observa-tion, combined with an estimate of the diameter of the Earth's orbit, yields for light a speed of about 220,000 kilome-ters a second.

Røemer's measurement was ac-cepted only when confirmed by James Bradley in 1729. Bradley's idea was to measure the speed of light relative to something else fast: the speed of the Earth as it orbited the Sun, which he knew. The Earth's motion changed the apparent direction at which light arrived from stars, just as your running through rain changes the angle at which rain-drops hit you. Bradley measured such shifts in position of stars and concluded that light travels at 298,000 kilometers a second. He was almost exactly right.

◄ A detailed view of Ganymede's Nicholson Regio shows a series of parallel scarp ridges, possibly the result of stretching and faulting of the crust.

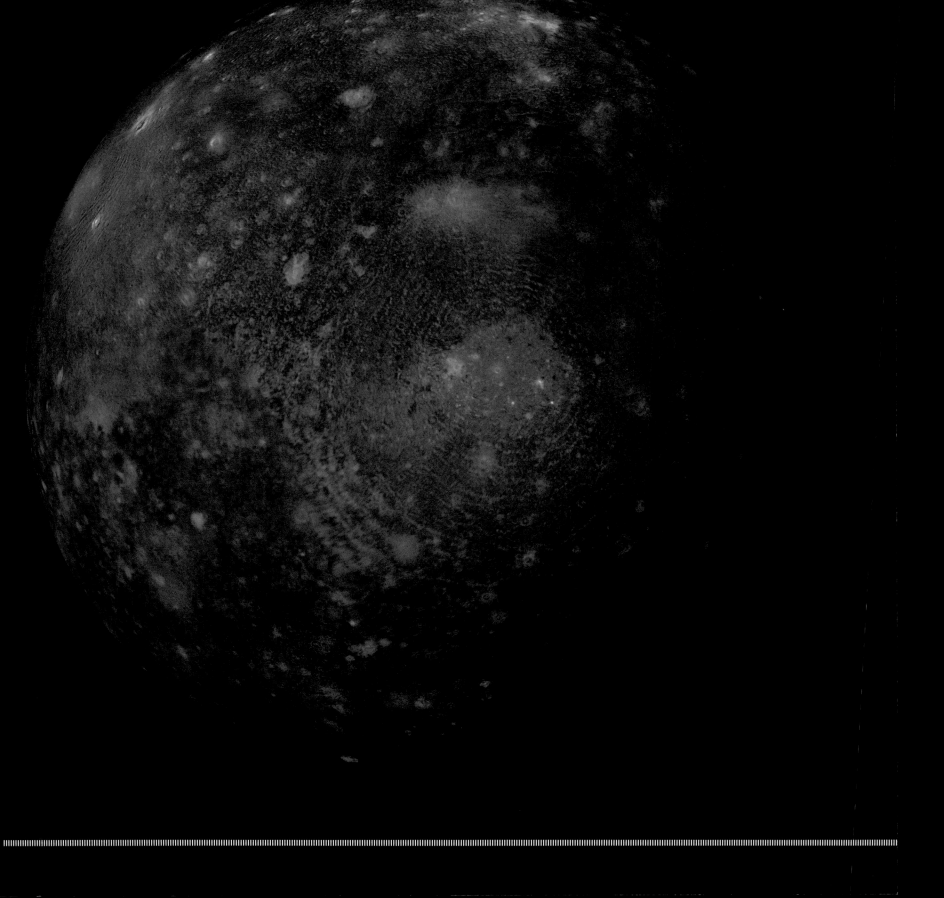

Callisto

ONE DAY, if humans survive environmental catastrophe, there will be a manned base on Callisto. Why? Because Callisto, the outermost of the four Galilean moons, is the only major satellite that orbits beyond Jupiter's deadly radiation belts. It is the safest place from which to explore the Jupiter system.

The third largest moon in the Solar System, after Ganymede and Saturn's Titan, Callisto is one of the most pockmarked worlds known. NASA's Galileo revealed that the moon's interior is mostly made of rock and ice. Like Europa, it may have a liquid ocean beneath its surface.

Along with the other Galilean moons, Callisto played a key role in the life of Italian scientist Galileo and his head-on collision with the Catholic Church.

▲ Impacts expose fresh bright ice in contrast to Callisto's dark, ancient surface. The craters Burr and Tormarsuk overlie the concentric rings of the older giant impact structure of Asgard.

▼ Callisto map based on images from the Galileo and Voyager missions. (Mollweide projection map centered on 0° longitude.)

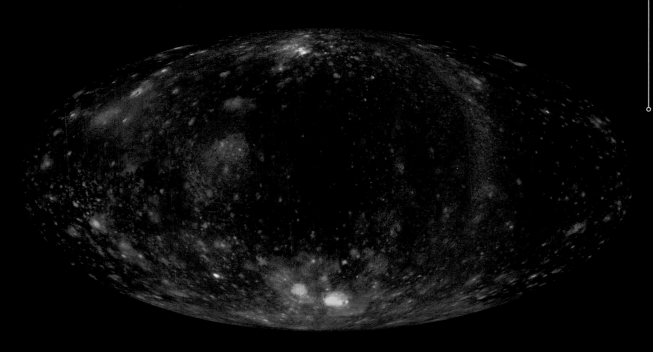

ORBITAL DATA
Distance from Jupiter 1,870,000 to 1,900,000 km
Orbital Period (Year) 16.69 Earth days
Length of Day 16.69 Earth days
Orbital Speed 8.3 to 8.1 km/s
Orbital Eccentricity 0.007
Orbital Incination 0.19°
Axial Tilt 0°

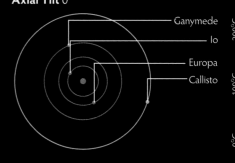

- Ganymede
- Io
- Europa
- Callisto

PHYSICAL DATA
Diameter 4,821 km / 0.38 x Earth
Mass 108 billion billion metric tons / 0.02 x Earth
Volume 58,700 million km³ / 0.05 x Earth
Gravity 0.126 x Earth
Escape Velocity 2.441 km/s
Surface Temperature 80° to 165°K / –193° to –108°C
Mean Density 1.851 g/cm³

The Moon

ATMOSPHERIC COMPOSITION
Carbon dioxide 99%
Oxygen 1%

Surface temperature

800 K
600 K
400 K
200 K
0 K
400°C
200°C
100°C
0°C

Mean density

Iron
Water
Rock

0
1g/cm³
2g/cm³
3g/cm³
4g/cm³
5g/cm³
6g/cm³
7g/cm³

Jupiter and the Church

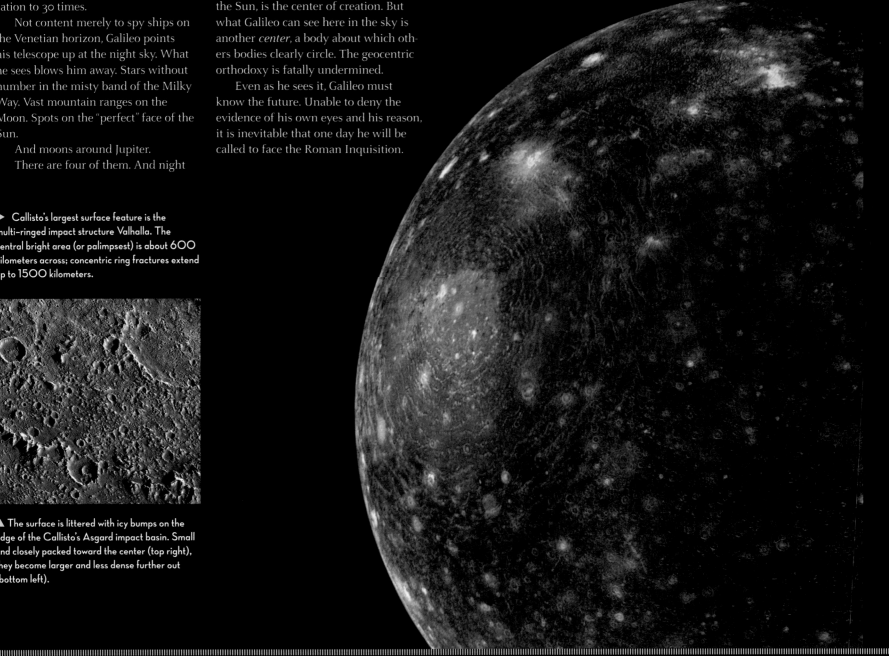

◀ Callisto's dark surface is uniformly cratered, but has color variations due to the varying mix of ice and rocky material on its surface.

IN JUNE 1609 Galileo Galilei hears of a new invention and drops everything. A Dutch lens maker called Hans Lippershey has put lenses at either end of a tube and, miraculously, made distant things appear close up. In a frenzy, Galileo builds his own "telescope," determines by experiment how to improve its performance, and boosts its magnification to 30 times.

Not content merely to spy ships on the Venetian horizon, Galileo points his telescope up at the night sky. What he sees blows him away. Stars without number in the misty band of the Milky Way. Vast mountain ranges on the Moon. Spots on the "perfect" face of the Sun.

And moons around Jupiter.

There are four of them. And night after night, as Galileo watches, they change their positions, whirling around the giant planet.

The significance of the discovery is huge. Seventy years earlier, the Polish astronomer Nicolaus Copernicus assembled powerful evidence that the planets circle the Sun. The Catholic Church maintains that the Earth, not the Sun, is the center of creation. But what Galileo can see here in the sky is another *center*, a body about which others bodies clearly circle. The geocentric orthodoxy is fatally undermined.

Even as he sees it, Galileo must know the future. Unable to deny the evidence of his own eyes and his reason, it is inevitable that one day he will be called to face the Roman Inquisition.

▶ Callisto's largest surface feature is the multi-ringed impact structure Valhalla. The central bright area (or palimpsest) is about 600 kilometers across; concentric ring fractures extend up to 1500 kilometers.

▲ The surface is littered with icy bumps on the edge of the Callisto's Asgard impact basin. Small and closely packed toward the center (top right), they become larger and less dense further out (bottom left).

AT THE LAST count, Jupiter had 63 moons. Leave aside the four giant Galilean moons and you have 59 left.

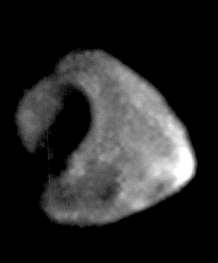
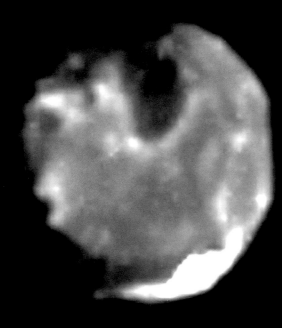
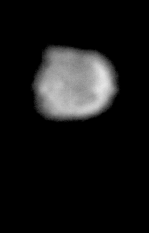

We are family

JUPITER'S BIGGEST "other moon" is Amalthea. A mere 168 kilometers across, it actually circles Jupiter within the orbit of Io, the innermost Galilean moon. Amalthea was the fifth moon to be discovered—in 1892, nearly 300 years after Galileo spotted the first four. It is one of four tiny inner moons, the others being Metis, Adrastea, and Thebe.

Adrastea and Metis orbit within the fine rings of dust that circle Jupiter. These flimsy disks, a pale shadow of Saturn's spectacular ring system, were discovered by NASA's Voyager 1 space probe in 1979.

By the time of the Voyager 1 flyby the number of known moons had risen to 16. Since then, another 47 have been found. Most are captured asteroids, mere pebbles less than 10 kilometers across. Captured moons, unlike home-grown ones, can have highly elongated orbits and travel in the opposite direction to their planet's spin.

The farthest Jovian moon, known rather unimaginatively as S/2003 J2, orbits at a distance of 29.5 million kilometers from the giant planet. This is about 75 times the distance of the Earth from the Moon, powerfully illustrating the immense reach of Jupiter's gravitational pull.

TOTAL NUMBER OF MOONS

63 (Metis, **Adrastea**, Amalthea, **Thebe**, Io, **Europa**, Ganymede, **Callisto**, Themisto, **Leda**, Himalia, **Lysithea**, Elara, **S/2000 J11**, S/2003 J12, **Carpo**, Euporie, **S/2003 J3**, S/2003 J18, **Orthosie**, Euanthe, **Harpalyke**, Praxidike, **Thyone**, S/2003 J16, **Iocaste**, Mneme, **Hermippe**, Thelxinoe, **Helike**, Ananke, **S/2003 J15**, Eurydome, **Arche**, Herse, **Pasithee**, S/2003 J10, **Chaldene**, Isonoe, **Erinome**, Kale, **Aitne**, Taygete, **S/2003 J9**, Carme, **Sponde**, Megaclite, **S/2003 J5**, S/2003 J19, **S/2003 J23**, Kalyke, **Kore**, Pasiphae, **Eukelade**, S/2003 J4, **Sinope**, Hegemone, **Aoede**, Kallichore, **Autonoe**, Callirrhoe, **Cyllene,** and S/2003 J2)

Saturn

TEN TIMES as far away from the Sun as the Earth and ten times the Earth's diameter, Saturn is the second largest planet in the Solar System. It has a unique and spectacular claim to fame. Nobody had the slightest suspicion of what that was until the invention of the telescope. Even then it took 50 years before anyone could make the slightest sense of what they were seeing.

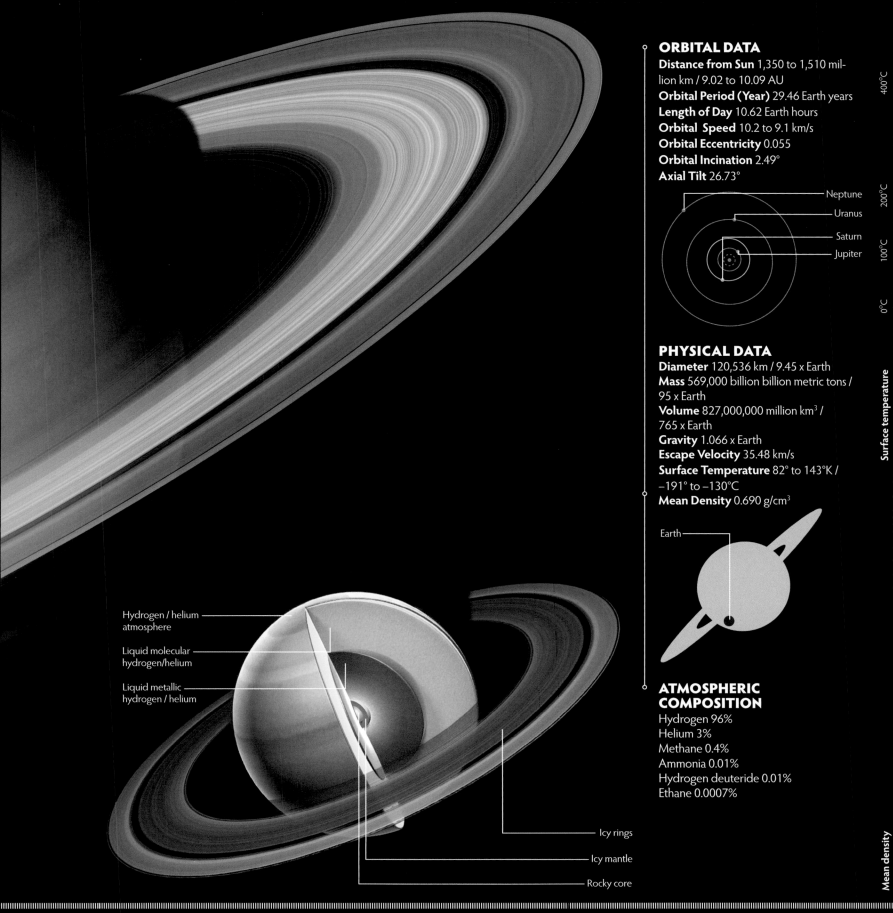

ORBITAL DATA

Distance from Sun 1,350 to 1,510 million km / 9.02 to 10.09 AU
Orbital Period (Year) 29.46 Earth years
Length of Day 10.62 Earth hours
Orbital Speed 10.2 to 9.1 km/s
Orbital Eccentricity 0.055
Orbital Incination 2.49°
Axial Tilt 26.73°

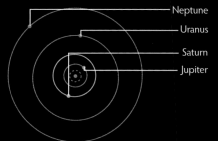

Neptune
Uranus
Saturn
Jupiter

PHYSICAL DATA

Diameter 120,536 km / 9.45 x Earth
Mass 569,000 billion billion metric tons / 95 x Earth
Volume 827,000,000 million km³ / 765 x Earth
Gravity 1.066 x Earth
Escape Velocity 35.48 km/s
Surface Temperature 82° to 143°K / –191° to –130°C
Mean Density 0.690 g/cm³

Earth

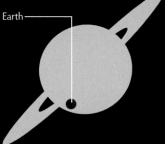

ATMOSPHERIC COMPOSITION

Hydrogen 96%
Helium 3%
Methane 0.4%
Ammonia 0.01%
Hydrogen deuteride 0.01%
Ethane 0.0007%

Hydrogen / helium atmosphere
Liquid molecular hydrogen/helium
Liquid metallic hydrogen / helium

Icy rings
Icy mantle
Rocky core

Surface temperature

800 K
600 K
400 K
200 K
0 K

400°C
200°C
100°C
0°C

Mean density

0
Water
Iron
Rock
1g/cm³
2g/cm³
3g/cm³
4g/cm³
5g/cm³
6g/cm³
7g/cm³

▶ Saturn map based on images from the Cassini mission. (Mollweide projection map centered on 0° longitude.)

▶ As Saturn's seasons advance, the color of the northern hemisphere changes to warmer hues, but still carries a trace of blue toward the pole in this image from 2008.

▶ This image combines three wavelengths of infrared light to examine heat emitted from Saturn. Blue and green show sunlight reflected at near-infrared wavelengths of 1 and 3 microns respectively; red shows heat emitted from Saturn's interior at 5 microns (thermal infrared).

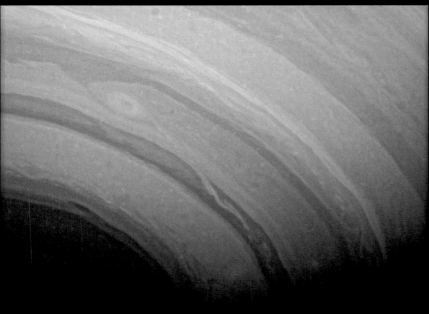

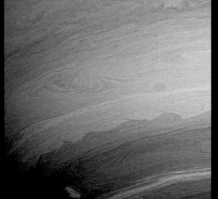

▲ A close-up infrared view of Saturn's southern hemisphere shows a highly dynamic cloudscape, in contrast to the bland, peaceful appearance in visible light.

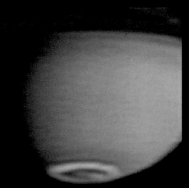

▲ Like Earth, Saturn exhibits polar aurora—the glow of atmospheric gases under the impact of the solar wind, which is funneled toward the poles by the planet's magnetic field.

▲ Saturn's latitudinal atmospheric bands are lit up in range of pastel colors when viewed at infrared wavelengths.

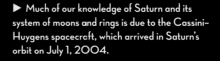

▶ Much of our knowledge of Saturn and its system of moons and rings is due to the Cassini–Huygens spacecraft, which arrived in Saturn's orbit on July 1, 2004.

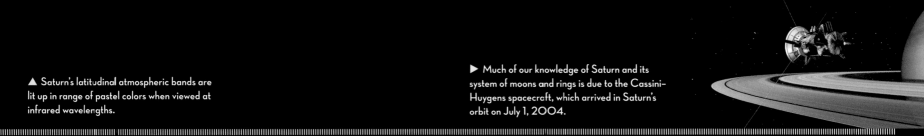

A planet with ears

GALILEO GALILEI is a giant in the history of science, discovering among other things that a swinging pendulum keeps perfect time. But one of his career low points was his claim that Saturn was a "planet with ears."

Unfortunately the telescope Galileo pointed at the night sky from Venice in 1610 was not powerful enough to reveal Saturn's big secret. At first he speculated that the planet had a moon on either side, each a third as big as the planet. However, in 1612 he was baffled when the two moons vanished. "Saturn has swallowed its children?" he wrote to his patron, the Grand Duke of Tuscany. In 1613 the moons reappeared, and Galileo was even more mystified.

The mystery was solved only in 1655, when Dutch scientist Christiaan Huygens built an improved telescope with a magnification of 50 and recognized, correctly, that Saturn was girdled by a wide system of rings. As the rings change their orientation, they can appear on either side of the planet, the "ears" seen by Galileo. Or they can appear edge-on, and vanish from view.

Today we know that the ring plane is tilted at 26.7 degrees to our line of sight. Although the orbiting rings keep their orientation in space, much like a spinning gyroscope, we on Earth see the rings at various angles as Saturn orbits the Sun. Twice during Saturn's 29.5-year orbit we see the rings edge-on.

The rings of Saturn are as familiar today as the McDonald's Golden Arches. They may even have inspired the bisected circle of the London Transport logo at every Tube station. But few people know that Saturn not only boasts rings but also has spots.

▶ The London Underground logo takes the form of Saturn with its rings.

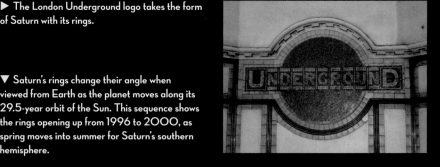

▼ Saturn's rings change their angle when viewed from Earth as the planet moves along its 29.5-year orbit of the Sun. This sequence shows the rings opening up from 1996 to 2000, as spring moves into summer for Saturn's southern hemisphere.

▶ Saturn's rings are apparent mainly as a dark shadow in this almost edge–on view a few months after the planet's equinox in 2009. The icy moon Rhea is visible at half phase in front of the planet, as is the shadow of the moon Tethys, on the left.

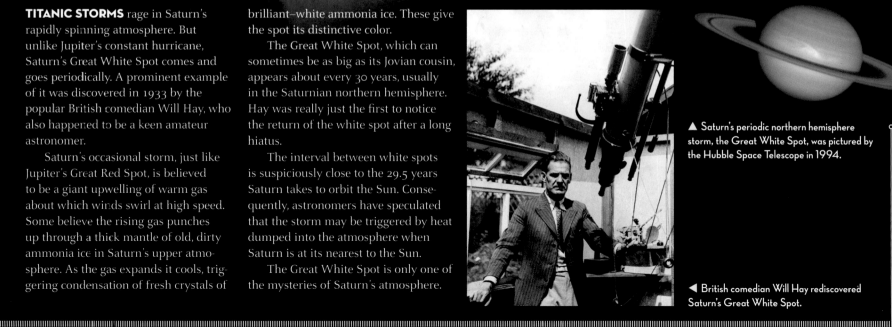

▶ Saturn's "Dragon Storm," a deep convective weather system with a complex shape, arose in September 2004. The Cassini Orbiter recorded powerful radio emissions from the area, similar to the static generated by lightning storms on Earth.

The Great White Spot

TITANIC STORMS rage in Saturn's rapidly spinning atmosphere. But unlike Jupiter's constant hurricane, Saturn's Great White Spot comes and goes periodically. A prominent example of it was discovered in 1933 by the popular British comedian Will Hay, who also happened to be a keen amateur astronomer.

Saturn's occasional storm, just like Jupiter's Great Red Spot, is believed to be a giant upwelling of warm gas about which winds swirl at high speed. Some believe the rising gas punches up through a thick mantle of old, dirty ammonia ice in Saturn's upper atmosphere. As the gas expands it cools, triggering condensation of fresh crystals of

brilliant–white ammonia ice. These give the spot its distinctive color.

The Great White Spot, which can sometimes be as big as its Jovian cousin, appears about every 30 years, usually in the Saturnian northern hemisphere. Hay was really just the first to notice the return of the white spot after a long hiatus.

The interval between white spots is suspiciously close to the 29.5 years Saturn takes to orbit the Sun. Consequently, astronomers have speculated that the storm may be triggered by heat dumped into the atmosphere when Saturn is at its nearest to the Sun.

The Great White Spot is only one of the mysteries of Saturn's atmosphere.

▲ Saturn's periodic northern hemisphere storm, the Great White Spot, was pictured by the Hubble Space Telescope in 1994.

◀ British comedian Will Hay rediscovered Saturn's Great White Spot.

Six appeal

WHEN "AIR" CIRCULATES in a planetary atmosphere, it circulates. . . well, in a circle. Whoever heard of a square hurricane? Or a hexagonal one? Yet this is exactly what has been found at one of the poles of Saturn.

In 2007, when the Cassini space probe flew over Saturn, it snapped the most bizarre image: a hexagonal arrangement of clouds turning around its north pole. The hexagon is almost twice as wide as the Earth. Peculiarly, Saturn's south pole has no matching hexagon, just clouds circulating around an "eye," as clouds do around the continent of Antarctica.

We know the honeycomb-shaped weather system is very stable and long-lived because it was first spotted by NASA's Voyager 1 and 2 space probes a quarter of a century earlier.

A clue to the origin of the polar hexagon comes from laboratory experiments in which a fluid is spun rapidly in a bucket. Under certain conditions, researchers have found that there appears spontaneously an unchanging or "standing" wave pattern in the shape of a polygon with three, four, five, or six sides. These geometric shapes are thought to be generated by the interaction between the fluid and the walls of the bucket.

The similarity between the fluid in the bucket and the atmosphere of the ringed planet strongly hints at a connection. But what could possibly be playing the role of the sides of the bucket?

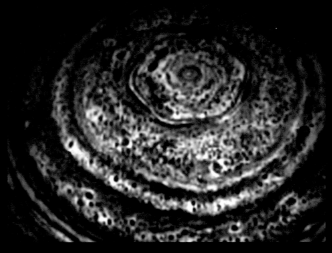

◄ Saturn's polar hexagon is revealed by this thermal image of the planet's north pole.

◄ Temperature patterns in Saturn's atmosphere follow the latitudinal bands seen at visible wavelengths, but there is a surprising hotspot at the north pole. Also highlighted by this temperature map looking down on Saturn's north pole is the circumpolar hexagon.

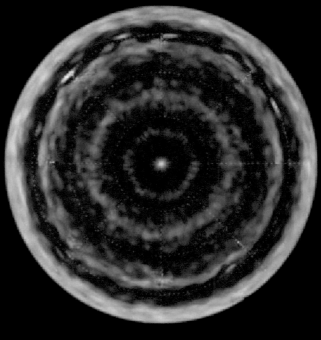

▲ An oblique view of the concentric walls of cloud comprising Saturn's intense south polar vortex.

▲ Staring into the eye of the hurricane at Saturn's south pole, walls of cloud can be seen towering 30 to 75 kilometers above the central area. The dark center of the storm is 8,000 kilometers across.

▼ Saturn's unique north polar hexagon is seen emerging from winter's shadow

◄ Saturn's rapid rotation and lightness cause its waist to bulge. The white line represents the change in the planet's dimensions.

Lightweight world

IMAGINE YOU HAVE a bowl of water big enough to hold all the planets. You throw them in, one at a time, and watch them sink like stones. Then you throw in Saturn. Alone among the planets, it floats. Although it looks solid it is only 70 percent as dense as water. Jupiter, by comparison, is a third denser than water.

We know this because there is a way of estimating how much mass there is in any given volume of a planet. First, Newton's law of gravity allows us to determine a planet's mass from the speed at which a moon is whirled around: the faster it orbits, the greater the mass of the planet. Estimating the volume of a planet is trickier. But by bouncing radar pulses off the planet and timing how long they take to return to Earth, it is possible to estimate the distance. This then allows the apparent size of the planet to be translated into its true size. Saturn is big enough to swallow 770 Earths.

Armed with an estimate of the amount of mass in a given volume it is possible to ask what the planet is made of. In the case of Saturn and Jupiter it can only be the lightest elements, hydrogen and helium. Saturn and Jupiter are both 90 percent hydrogen and 10 percent helium atoms. This reflects the makeup of the primordial nebula from which the Solar System formed.

Because Saturn is so insubstantial and spins so fast–once every 10 hours or so–it bulges at its equator by 11 percent, more than any other planet.

Saturn's Rings

THREE OTHER PLANETS in the Solar System have rings, but none bear comparison with the rings of Saturn. They are so impossibly thin they vanish entirely when seen edge-on. Yet if they were around the Earth, they would stretch more than a third of the way to the Moon. What are they made of?

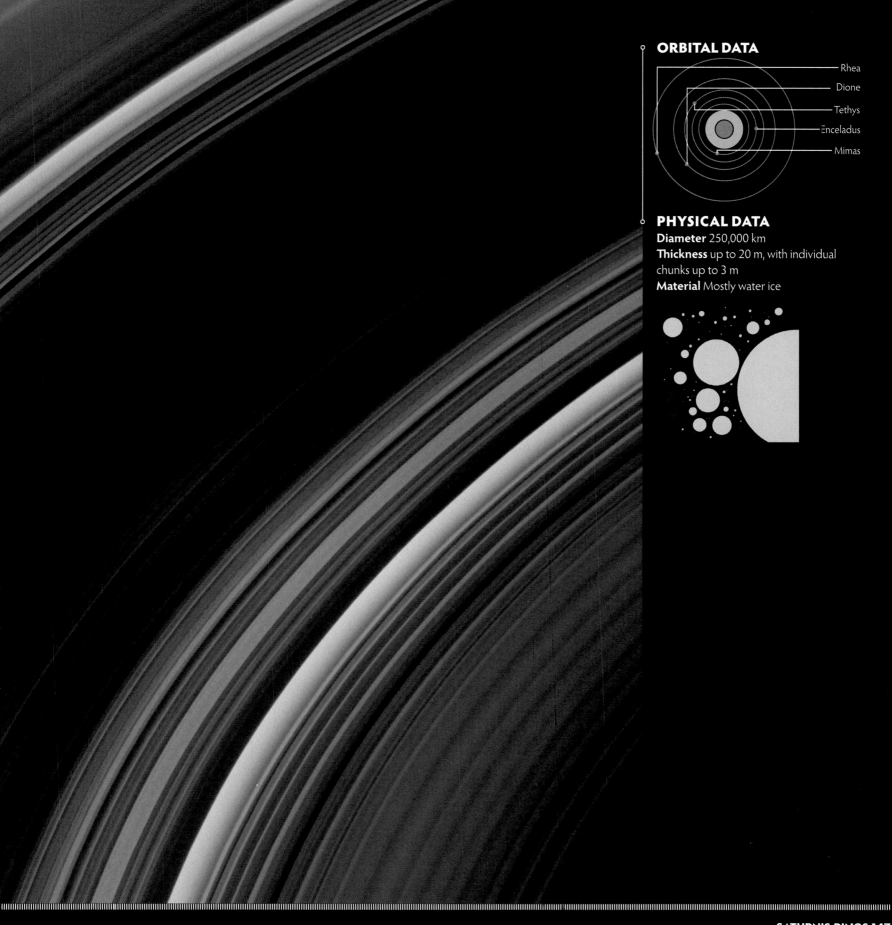

Rhea
Dione
Tethys
Enceladus
Mimas

PHYSICAL DATA

Diameter 250,000 km
Thickness up to 20 m, with individual chunks up to 3 m
Material Mostly water ice

◀ Infrared imagery shows the outer G and E rings and the inner D ring in blue and green, indicating they are composed of small particles relative to the main rings, which show as yellow and red.

▶ Voyager 2 discovered "spokes" in Saturn's B ring– dark radial features that rotated with the ring system.

▼ Waves, gaps, resonance patterns, and subtle color variations are shown in this detailed pan-orama across Saturn's main ring system.

Colombo Gap

Maxwell Gap

C Ring

B Ring

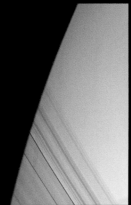

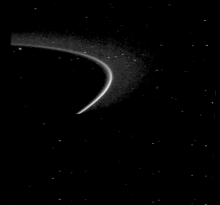

▲ The delicate strands of Saturn's inner C ring, seen here from the unilluminated side, draw a thin veil across the face of Saturn.

▲ A long-exposure time shows details in the unilluminated side of Saturn's rings, at the expense of over-exposing the planet itself.

▲ A bright arc in Saturn's diffuse G ring is shown here disappearing into the planet's shadow.

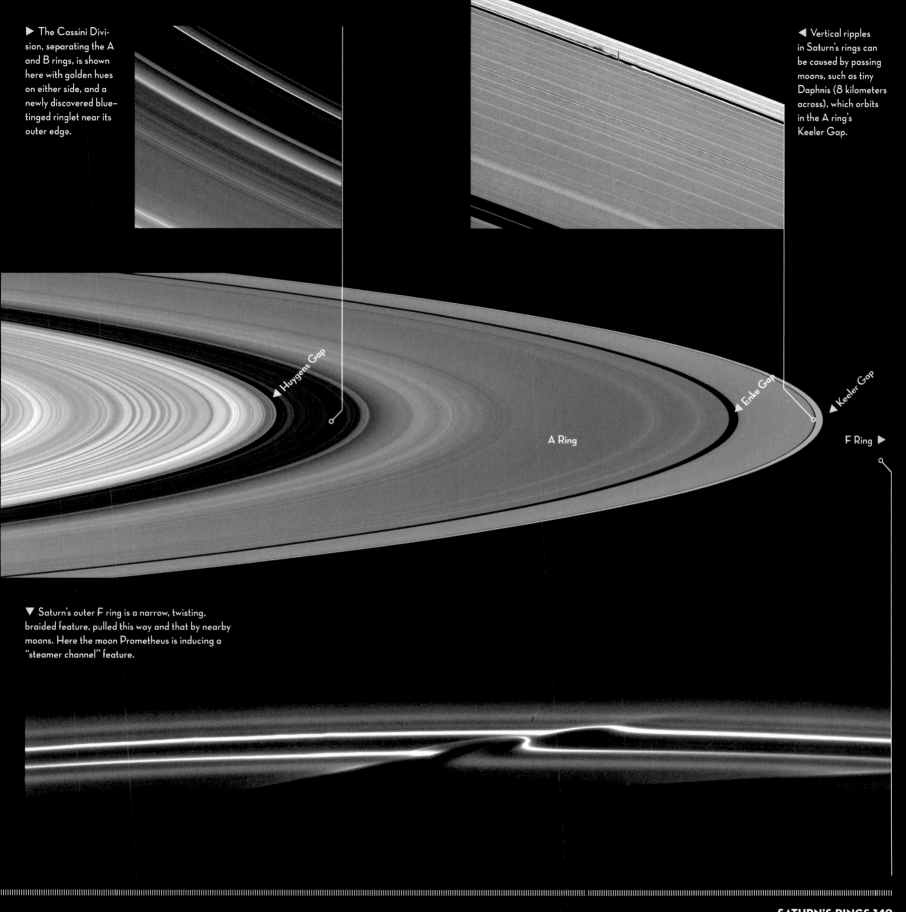

▶ The Cassini Division, separating the A and B rings, is shown here with golden hues on either side, and a newly discovered blue–tinged ringlet near its outer edge.

◀ Vertical ripples in Saturn's rings can be caused by passing moons, such as tiny Daphnis (8 kilometers across), which orbits in the A ring's Keeler Gap.

Huygens Gap

Enke Gap

Keeler Gap

A Ring

F Ring ▶

▼ Saturn's outer F ring is a narrow, twisting, braided feature, pulled this way and that by nearby moons. Here the moon Prometheus is inducing a "steamer channel" feature.

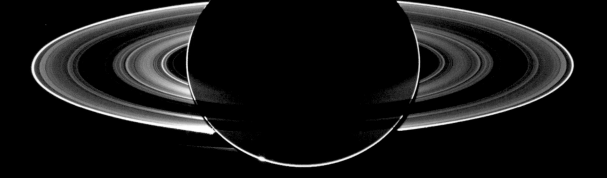

▲ In this Cassini image, taken looking back toward the sun, the rings are revealed in all their glory. Recognize the pale blue dot on the left, between the two faint outer rings? It's the Earth.

▲ The colors here are related to the size of the particles in Saturn's rings. Boulder-sized particles several meters across are common throughout, but purple shows a lack of smaller particles less than 5 centimeters across. Green shows particles smaller than 5 centimeters, and blue smaller than 1 centimeters.

JAMES CLERK MAXWELL was the greatest physicist between the time of Newton and Einstein. He summarized all electrical and magnetic phenomena in one set of equations, discovering that light was a wave of electromagnetism. Before all this, he tackled the problem of Saturn's rings.

The key question was: are they solid, fluid, or composed of many separate particles? In 1858, in a mathematical tour de force, Maxwell showed that if the rings were solid or fluid-like they would disintegrate. The rings, he concluded in an essay which won the prestigious Adams Prize, must be made of countless particles sweeping around Saturn like a swarm of micro-moons.

Maxwell would have been pleased by the images sent back by NASA's Voyager 1 and 2 space probes in 1980 and 1981. Although from Earth astronomers see only a handful of wide rings interrupted by gaps, the Voyagers revealed tens of thousands of narrow ringlets. The inner ringlets rotate faster than the outer ones, confirming Maxwell's conclusion that they could never be solid. (Strictly speaking, Saturn's rings are actually multiple spirals, each like the groove on an old vinyl record. Long known by researchers in the field, this fact has not yet become well known to the public.)

The rings are made of countless chunks of 99 percent water-ice sparkling in the sunlight, which is why they appear so bright. They range in size from smaller than a sand grain to as big as a two-story house. The brightest rings must be made of particles that are highly reflective, probably because they are fluffy snowballs whose shape gives them a relatively large surface area. All the ring particles orbit in a layer probably less than about 20 meters thick. If the rings were shrunk to a disk a kilometer across, they would be thinner than the sharpest razor blade.

Genesis of the rings

IF YOU PUT ALL the ring particles together they would form an object about 200 to 300 kilometers across, the size of a medium–sized moon of Saturn. This may be a clue to the origin of the rings.

The 19th–century French mathematician Édouard Roche speculated that once upon a time a moon drifted too close to Saturn. Once inside the so-called "Roche limit," differences in the strength of Saturn's gravity across the width of the moon were big enough to tear the moon apart. Another possibility is that the rings are the remains of a giant comet that disintegrated when it flew too close to Saturn.

The gravity between Saturn's small moons and its ring particles in effect causes them to repel each other: while the moons move gradually outward, the ring particles move inward. The best estimate is that the rings will fall into Saturn in about 400 million years. Are we just lucky to be around to see them?

Luck may not be necessary. The rings could be far older if, as fast as they lose material, the moons are resupplying them with ice. (The moon Enceladus is doing just this.) The problem then is to answer the question: why do the rings look young? They are white and pristine, but dust from meteorite impacts over the ages should have darkened them.

There is a way out of this dilemma. The ring material is continually clumping together and being shattered by meteorite impacts. This recycling, akin to breaking open a snowball to reveal pristine ice, may be making the rings appear youthful even if they are not.

▼ A comparison of the ring systems of the outer planets, scaled to a common radius. In each case, most ring material lies within the planet's Roche limit.

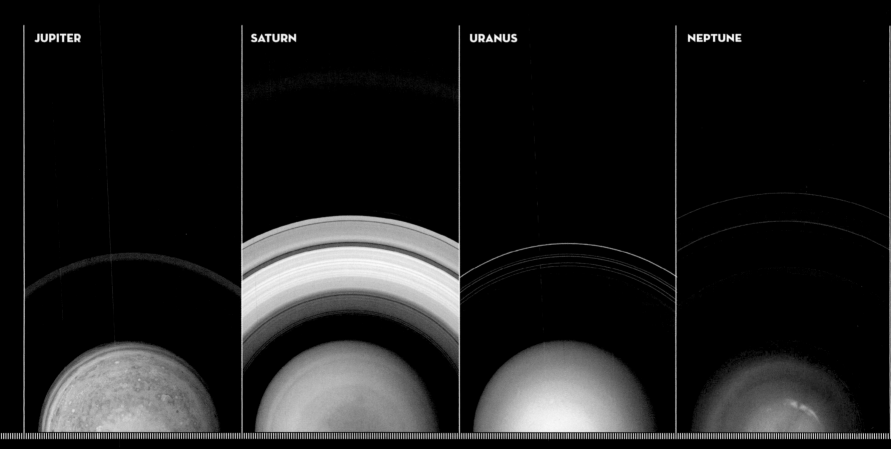

JUPITER

SATURN

URANUS

NEPTUNE

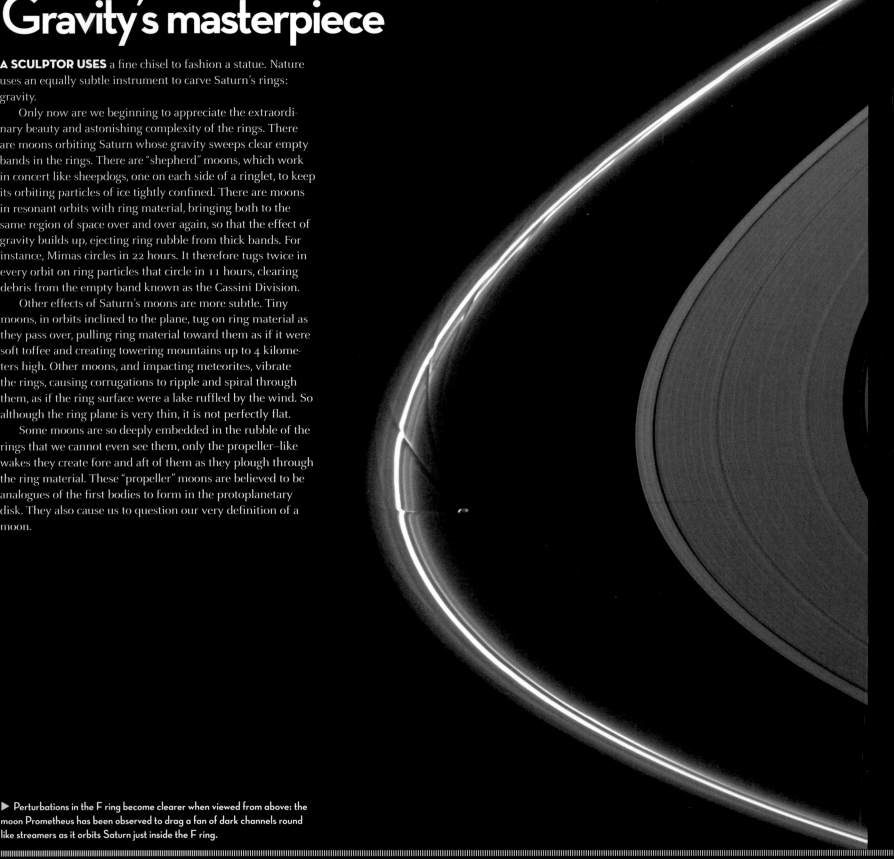

Gravity's masterpiece

A SCULPTOR USES a fine chisel to fashion a statue. Nature uses an equally subtle instrument to carve Saturn's rings: gravity.

Only now are we beginning to appreciate the extraordinary beauty and astonishing complexity of the rings. There are moons orbiting Saturn whose gravity sweeps clear empty bands in the rings. There are "shepherd" moons, which work in concert like sheepdogs, one on each side of a ringlet, to keep its orbiting particles of ice tightly confined. There are moons in resonant orbits with ring material, bringing both to the same region of space over and over again, so that the effect of gravity builds up, ejecting ring rubble from thick bands. For instance, Mimas circles in 22 hours. It therefore tugs twice in every orbit on ring particles that circle in 11 hours, clearing debris from the empty band known as the Cassini Division.

Other effects of Saturn's moons are more subtle. Tiny moons, in orbits inclined to the plane, tug on ring material as they pass over, pulling ring material toward them as if it were soft toffee and creating towering mountains up to 4 kilometers high. Other moons, and impacting meteorites, vibrate the rings, causing corrugations to ripple and spiral through them, as if the ring surface were a lake ruffled by the wind. So although the ring plane is very thin, it is not perfectly flat.

Some moons are so deeply embedded in the rubble of the rings that we cannot even see them, only the propeller–like wakes they create fore and aft of them as they plough through the ring material. These "propeller" moons are believed to be analogues of the first bodies to form in the protoplanetary disk. They also cause us to question our very definition of a moon.

▶ Perturbations in the F ring become clearer when viewed from above: the moon Prometheus has been observed to drag a fan of dark channels round like streamers as it orbits Saturn just inside the F ring.

A MOON IS a small body orbiting a big body. This definition enables a body like Titan, which is bigger than a planet like Mercury, to be classified as a moon. (Let's put aside binary asteroids, where the two bodies in orbit about each other are the same size!) We can all agree that a moon orbits in empty space, is long-lived, and of a reasonable size, right? The only problem is that many of Saturn's moons do not orbit in empty space, they are embedded in its ring material.

Imagine that you could zoom in on an individual ringlet. Rather than an unbroken line of material, you would see a broken, dashed line. Ring particles are continually piling up, sticking together, then coming apart. These ephemeral bodies, which never reach more than 30 to 50 meters across, hover on the boundary between a moon and a collection of debris. Saturn is a laboratory for determining how small a body can be and still be considered a moon.

Saturn's rings are a model of the protoplanetary disk from 4.55 billion years ago. When we observe moons coming together out of the debris, some only briefly but others surviving for several orbits before disappearing, it is as if we are observing the complex processes that eventually led to the birth of the Earth and planets.

▲ The "clumping" of ring particles is shown by a gravitational simulation produced by Heikki Salo at University of Oulu, Finland.

Titan

TITAN IS THE SECOND biggest moon in the Solar System. Like Jupiter's Ganymede, it is bigger than the planet Mercury. A source of fascination for planetary scientists and biologists alike, it is the only body in the outer Solar System on which a space probe has landed.

▼ Titan surface map based on infrared images from the Cassini mission. (Mollweide projection map centered on longitude 180° West.)

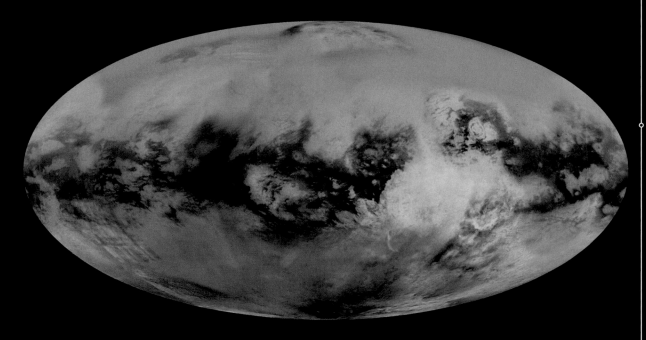

▶ Titan emerges from behind Saturn. The rings appear unusually dark, since we're seeing the shadowed side from the Cassini spacecraft.

ORBITAL DATA

Distance from Saturn 1,180,000 to 1,250,000 km
Orbital Period (Year) 15.88 Earth days
Length of Day 15.95 Earth days
Orbital Speed 5.8 to 5.4 km/s
Orbital Eccentricity 0.0292
Orbital Incination 0.35°
Axial Tilt 0°

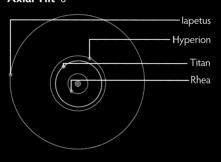

Iapetus
Hyperion
Titan
Rhea

PHYSICAL DATA

Diameter 5,150 km / 0.40 x Earth
Mass 135 billion billion tons metric / 0.02 x Earth
Volume 71,500 million km^3 / 0.07 x Earth
Gravity 0.139 x Earth
Escape Velocity 2.645 km/s
Surface Temperature 94 K / −179°C
Mean Density 1.881 g/cm^3

The Moon

ATMOSPHERIC COMPOSITION

Nitrogen 98.4%
Methane 1.4%
Hydrogen 0.2%

Surface temperature

400°C
200°C
100°C
0°C

800 K
600 K
400 K
200 K
0 K

Mean density

0
1g/cm^3
2g/cm^3
3g/cm^3
4g/cm^3
5g/cm^3
6g/cm^3
7g/cm^3

Water
Rock
Iron

A world like our own

IT IS JANUARY 2005. Parachuting through the thick orange haze of Titan's atmosphere, the Huygens lander radios back pictures to the Cassini mother ship in orbit around Saturn. The pictures reveal startlingly familiar features: rivers running down hillsides and emptying into an ocean along a jagged coastline.

The Huygens lander, built by the European Space Agency, comes down with a bump. Nobody knew whether it would strike solid ground or liquid, so it has been designed to soft-land and float. Its cameras pan around the alien landscape. But the landscape does not appear so alien. The lander has come to rest in the middle of a field of perfectly smooth, shiny white boulders, formed not of rock but of rock–hard ice.

It is exactly like a river delta on Earth. The boulders have been deposited by gushing liquid. They are smooth because they have been repeatedly smashed and jounced together in a torrent, roaring and foaming as it cascaded down to the sea.

Here is striking evidence of a world like our own. Except that it is 180˚ below the freezing point of water.

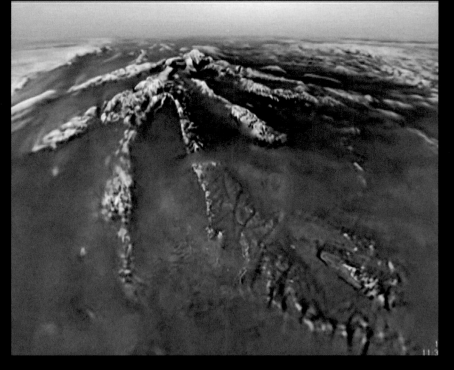

▲ View of Titan's surface from ESA's Huygens lander (monochrome image with simulated color).

▼ The Huygens Lander touched down on the surface of Titan on January 14th, 2005, after the descent of 2 hours 28 minutes. It continued to transmit data from the surface for another two hours.

▶ The dark equatorial band seen in infrared images is shown to be covered by longitudinal sand dunes in detailed radar images. The dunes are formed by winds blowing from the west and northwest.

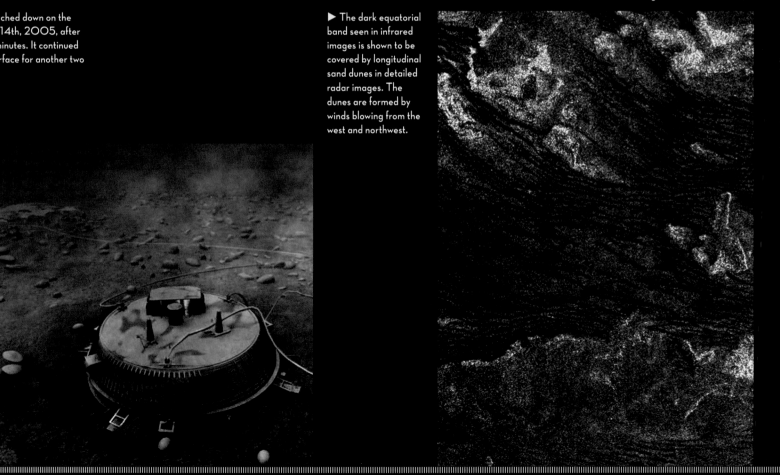

Windsurfing on lighter fluid

IN THE GRAINY half–light, beneath a sky thick with orange clouds, a stream gurgles its way down a rocky mountainside. Only it is not water doing the gurgling–it is far too cold for that. It is lighter fluid. On Titan water has been usurped by a mixture of liquid methane and ethane.

Earth is the only place in the Solar System where water exists in all three phases: as a liquid, a vapor, and a solid. On Titan other substances have stepped up to the plate. Methane and ethane are present here as a gas, liquid, and solid. Water, meanwhile, is frozen as hard as steel.

On Titan, there are oceans and lakes of liquid methane and ethane lighter fluid. When the liquid evaporates, it falls as rain and snow–giant flakes of wax spinning languidly groundward in the moon's dense atmosphere and low gravity. Rivers and streams return liquid to the oceans. Titan is the only world other than Earth with a complex "water cycle"–not with water but a water substitute.

Oddly, Titan's oceans have no waves to speak of. Radio waves bounced off the oceans by the Cassini space probe reveal they are totally smooth. The surface of Ontario Lacus, in the northern hemisphere, varies by less than a few millimeters over 100 meters. This is believed to be because a liquid methane-ethane mix is more viscous than water and the winds on super-cold Titan are too weak to ruffle it. Windsurfing on Titan would actually be a dead loss.

◀ This map centered on Titan's north pole shows the strips of ground that have been imaged by Cassini's radar. About 14 percent of the imaged area is radar–dark, interpreted as having a surface of liquid hydrocarbons.

▶ Lakes of lighter fluid: liquid shows up as blue in this image of Titan created by radar echoes picked up by Cassini.

Earth in the deep freeze

ON NOVEMBER 12, 1980, when Voyager 1 flew through the Saturnian system and pointed its camera at the giant moon, there was disappointment and jubilation. Disappointment because the surface of Saturn's biggest moon was entirely hidden from view. Jubilation because it was hidden from view by one of the thickest, most mysterious atmospheres in the Solar System.

A few big moons have atmospheres, but they are really just thin veils of gas. Titan's, though, is about four times as dense as the Earth's and exerts about one and a half times the surface pressure.

That atmosphere is principally composed of nitrogen, the gas that makes up roughly 80 percent of the Earth's atmosphere. And it is topped by a thick orange haze–a photochemical smog pretty much like that which builds up over Los Angeles.

The smog is a witch's cauldron of hydrocarbons evaporated from the oceans. Chemical reactions energized by feeble sunlight may create the building blocks of DNA such as amino acids that then drizzle down out of the smog, coating the surface with a sticky goo.

With its nitrogen atmosphere and soup of biochemicals, Titan is like a primordial Earth, preserved for 4.55 billion years in interplanetary deep freeze. The question must be: is there any life there?

▼ Titan's thick atmosphere presents a bland featureless mask at visible wavelengths in the image at left. Titan's atmosphere is more transparent at infrared wavelengths in the center image, allowing us to see dark and bright areas on the surface. The image on the right presents a composite of three infrared images and helps to sharpen the detail seen on Titan's surface.

▲ In Titan's blue upper atmosphere, methane is broken apart by ultraviolet light from molecules such as ethane and acetylene. The orange atmosphere below is a thick smog of complex organic molecules, which allows only 10% of the sunlight to reach the surface.

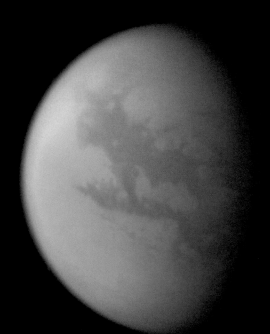

Life on Titan?

IN JAMES HOGAN'S 1983 novel *Code of the Lifemaker*, an alien spaceship crash-lands on Titan a million years before humanity. Its malfunctioning systems seed the moon with imperfect machine organisms, which begin rapidly to evolve. By the time humans reach it in the 21st century, a civilization has already developed on Titan.

The irony is that Titan already has all the ingredients needed for the emergence and evolution of life. The only problem is the mind-numbingly low temperature. At minus 180°C, chemical reactions would proceed at a snail's pace. Titan may be on its way to developing a biosphere, but without the warmth to kick-start the process it might take longer than the present age of the Universe.

However, things could change dramatically in a few billion years' time when the Sun runs out of its hydrogen fuel and swells into a red giant, pumping out 10,000 times as much heat as it does now. Lifted from its agelong deep freeze, Titan could become a paradise for life.

▲ Ontario Lacus is the largest lake on Titan's southern hemisphere, about 200 km long by 90 km wide. A two-lobed river delta can be seen on its eastern shoreline.

▶ Spectral measurements taken on Titan's surface were used to add color to the monochrome view of the Huygens lander's cameras.

Enceladus

A MERE 498 KILOMETERS across, Enceladus is pretty much the same size as Mimas. But here the similarity ends. Whereas Mimas is a crater-strewn and dead moon, Enceladus, to everyone's amazement, is alive and kicking.

▼ Enceladus map based on images from the Cassini mission. (Mollweide projection map centered on longitude 90° East.)

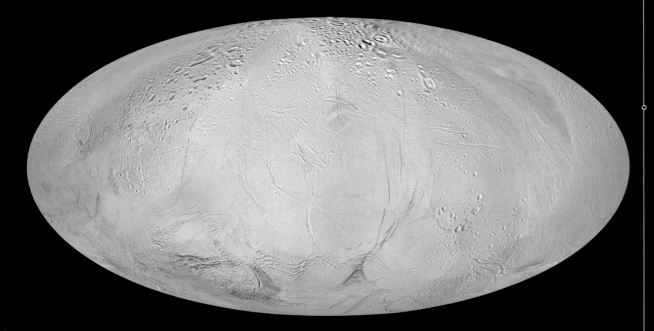

◄ Hanging just above the ring plane, Enceladus shines brightly as night falls on Saturn.

ORBITAL DATA
Distance from Sun 237,000 to 239,000 km
Orbital Period (Year) 1.37 Earth days
Length of Day 1.37 Earth days
Orbital Speed 12.7 to 12.6 km/s
Orbital Eccentricity 0.0045
Orbital Incination 0.02°
Axial Tilt 0°

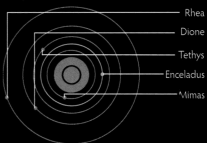

Rhea
Dione
Tethys
Enceladus
Mimas

PHYSICAL DATA
Diameter 500 km / 0.04 x Earth
Mass 110,000 million million metric tons
Volume 66 million km^3
Gravity 0.012 x Earth
Escape Velocity 0.242 km/s
Surface Temperature 33 to 145 K / −240° to −128°C
Mean Density 1.120 g/cm^3

Ireland

ATMOSPHERIC COMPOSITION
Water 91%
Nitrogen 4%
Carbon dioxide 3.2%
Methane 1.7%

Surface temperature

400°C
200°C
100°C
0°C

800 K
600 K
400 K
200 K
0 K

Mean density

Water
Rock
Iron

1g/cm³
2g/cm³
3g/cm³
4g/cm³
5g/cm³
6g/cm³
7g/cm³

The ice fountains of Enceladus

IT IS ONE of the most surprising and extraordinary images in the history of planetary exploration. Spouting from tiny Enceladus, illuminated by the Sun and extending hundreds of kilometers into space, are gargantuan fountains of sparkling ice crystals.

Before NASA's Cassini space probe took this startling image in November 2008, there had been hints that Enceladus is not a dead moon. It is the whitest, shiniest body in the Solar System. The dirt and grime it should have accumulated over the ages has been overlain by fresh snow. And in the moon's southern hemisphere there are four mint–colored fractures, dubbed tiger stripes, indicating surface movement and therefore warmth. No scientist was surprised to learn that this region is the source of Enceladus's ice fountains.

So much activity on such a tiny and cold body is a huge surprise. Some of the heat that drives the fountains may come from tidal stretching of the moon by Dione, which orbits Saturn once for every two orbits made by Enceladus. Still, there must be another, as yet unknown, heat source. Enceladus can eject ice crystals at more than 2,000 kilometers an hour, a speed attained in terrestrial geysers only when hot water under pressure is involved. Enceladus joins a select group of bodies–basically, Mars and Europa–suspected of containing water. Incredibly, beneath the ice crust of Enceladus there may be a global ocean.

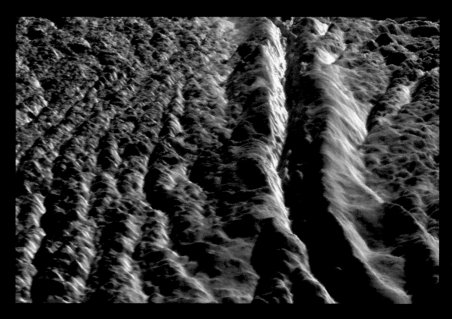

▲ Damascus Sulcus is one of the prominent "tiger stripes"; identified as the source of Enceladus's polar jets. It is a 5-kilometer-wide, 140-kilometer-long V–shaped valley with ridges rising to 250 meters at either side.

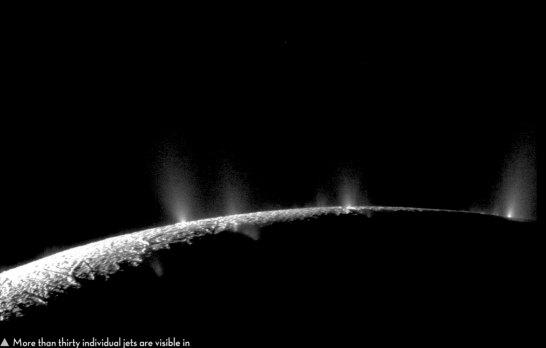

▲ More than thirty individual jets are visible in this detailed view of the south polar region of Enceladus.

▲ Enceladus leaks plumes of water ice from a region near its south pole, seen here dramatically back-lit by the Sun. Enceladus itself is illuminated by light reflected from Saturn.

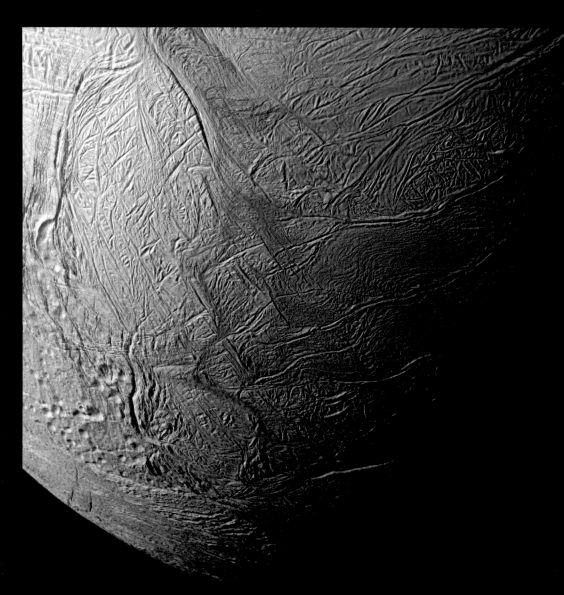

The smallest ocean

AN OCEAN on a body barely the size of England or Arizona beggars belief.

Once we thought liquid water was possible only in a narrow "habitable zone" around the Sun. Then it was realized that the tidal stretching of a moon by the gravity of a planet could heat its interior, even if the sunlight it was receiving were feeble. But surely such an effect would be appreciable only on a big moon. Nobody imagined it in anything as small as Enceladus.

Saturn's tiny ice moon has turned our ideas about where we might find life on their head. Deep in its interior are all the ingredients necessary for biology—water, warmth, and organic molecules. (Organic molecules are known to be present because they are responsible for the mint-green color of the moon's tiger stripes.)

On Earth, colonies of organisms thrive in the complete absence of sunlight around sub-sea volcanic vents, gushing hot water and chemicals. It is possible to imagine similar ecosystems on Enceladus. Who knows what might be inside the tiny moon? Is there a teeming ecosystem of microorganisms, hidden away in darkness since the birth of the Solar System? If so, Saturn's E ring, fed by Enceladus's ice fountains, may be more than simply water-ice. It may be an orbital graveyard of frozen microorganisms.

▲ Subtle color variations across the surface of Enceladus are revealed by using light extending beyond the visible range into the infrared and ultraviolet parts of the spectrum.

Iapetus

IAPETUS WAS discovered by the Italian-French astronomer Giovanni Cassini in 1672. Of all the bodies in the Solar System, it is the most two-faced.

▼ Iapetus map based on images from the Cassini and Voyager 2 missions. (Mollweide projection map centered on longitude 180° West.)

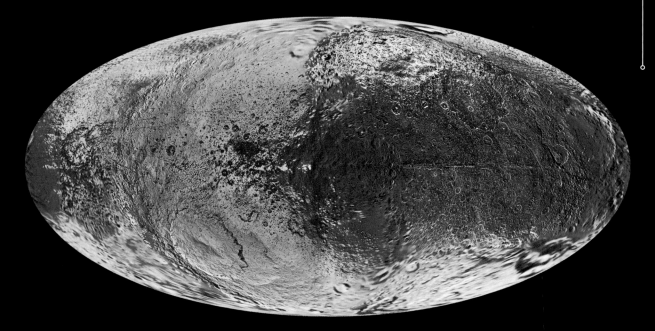

ORBITAL DATA

Distance from Saturn 3,460,000 to 3,660,000 km
Orbital Period (Year) 79.35 Earth days
Length of Day 79.35 Earth days
Orbital Speed 3.4 to 3.2 km/s
Orbital Eccentricity 0.0286
Orbital Incination 15.47°
Axial Tilt 0°

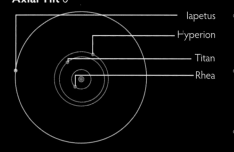

- Iapetus
- Hyperion
- Titan
- Rhea

PHYSICAL DATA

Diameter 1,470 km / 0.12 x Earth
Mass 1.8 billion billion metric tons
Gravity 0.023 x Earth
Escape Velocity 0.572 km/s
Surface Temperature 100 to 130 K / –173° to –143°C
Mean Density 1.020 g/cm^3

ATMOSPHERIC COMPOSITION

Hydrogen 96%
Helium 3%
Methane 0.4%
Ammonia 0.01%
Hydrogen deuteride 0.01%
Ethane 0.0007%

Surface temperature

800 K — 400°C
600 K
400 K — 200°C
200 K — 100°C
0°C
0 K

Mean density

0
Water — 1g/cm^3
2g/cm^3
Rock — 3g/cm^3
4g/cm^3
5g/cm^3
6g/cm^3
Iron — 7g/cm^3

Stargate moon

IAPETUS PLAYS a central role in Arthur C. Clarke's novel *2001: A Space Odyssey*. It is the site of the star gate moon, the portal through which Dave Bowman–Clarke's modern-day Odysseus–journeys to meet his destiny across the Universe.

Clarke selected Iapetus because, mysteriously, it has one face 10 times brighter than the other. What better place to locate an artificial alien artifact than on a moon that appears to be artificial?

The key to understanding Iapetus's Janus faces is an extraordinary ridge, stretching almost a third of the way around its equator and in places more than twice the height of Everest. The ridge bisects the dark side.

According to Paulo Freire of the Arecibo Observatory in Puerto Rico, Iapetus may have grazed Saturn's rings pretty much like a rock encountering a trimmer. Touching the rings for just three hours would be sufficient to create a ridge 5 kilometers high and 10 kilometers wide.

Freire says the rings' tiny dust grains are a perfect source of blackening material. They also contain ice–such as carbon dioxide, or dry ice, which in such a collision would instantly vaporize. Superfast winds would blow out from the ridge to deposit dust over a large area and blacken the face centered on the ridge.

If Iapetus did hit Saturn's rings, it must have been orbiting in the same plane as the rings, as the ridge lies along its equator. Since Iapetus no longer orbits in this plane, Freire speculates it must somehow have been ejected into its current orbit, perhaps by colliding with another moon.

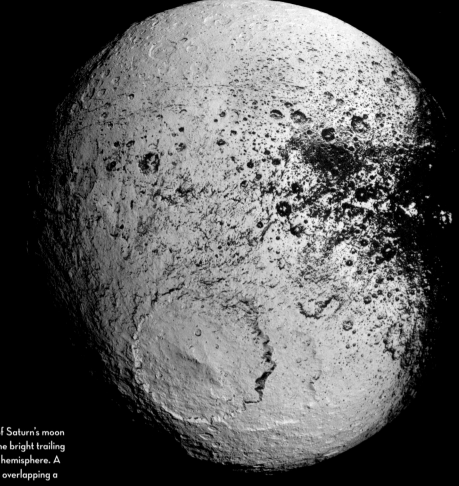

▶ The most noticeable feature of Saturn's moon Iapetus is the contrast between the bright trailing hemisphere and the dark leading hemisphere. A bedrock of bright ice seems to be overlapping a loose coating of dark material.

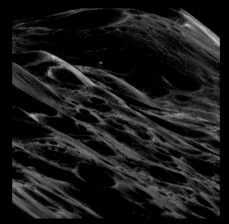

▲ A close up of the equatorial ridge of Iapetus shows mountains reaching 10,000 meters in height.

▶ Toward the north pole of Iapetus, the dark material coats the warmer south-facing walls of the craters. The coating does not seem to survive so well in colder regions.

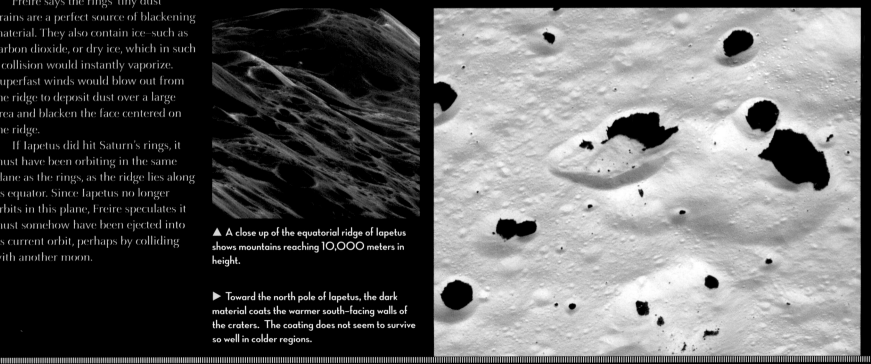

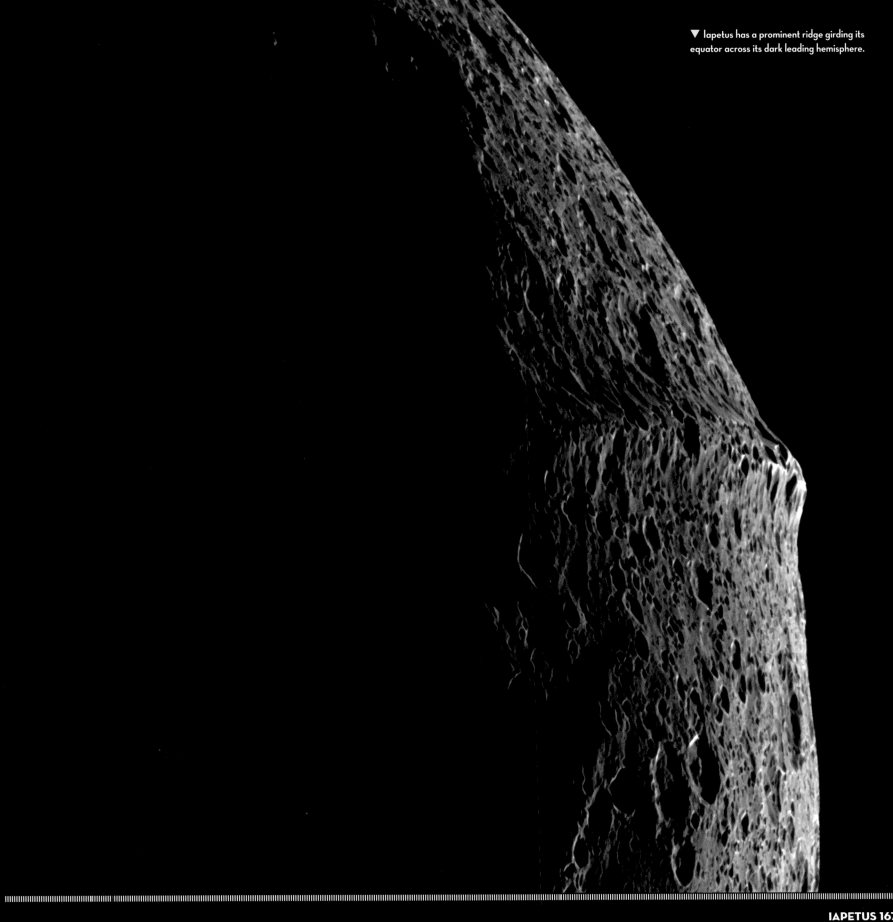

▼ Iapetus has a prominent ridge girding its equator across its dark leading hemisphere.

Mimas

ITS FACE IS SCARRED by a crater a third of the diameter of the moon itself. Imagine a crater on the Earth as wide as the Atlantic Ocean. Mimas, a tiny ball of rock and ice, is often called the Death Star moon on account of its resemblance to the moon-sized super-weapon in *Star Wars*. The tiny world's mega-crater is comparable to Stickney on Phobos and the Imbrium Basin on the Moon. It is a wonder that Mimas could have taken such a big hit and not disintegrated altogether. How did it survive?

ORBITAL DATA
Distance from Saturn 182,000 to 189,000 km
Orbital Period (Year) 0.94 Earth days
Length of Day 0.94 Earth days
Orbital Speed 14.6 to 14.0 km/s
Orbital Eccentricity 0.0202
Orbital Incination 1.57°
Axial Tilt 0°

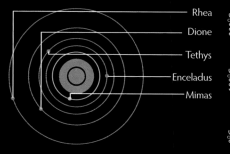

- Rhea
- Dione
- Tethys
- Enceladus
- Mimas

PHYSICAL DATA
Diameter 400 km / 0.03 x Earth
Mass 40,000 million million metric tons
Volume 34 million km³
Gravity 0.007 x Earth
Escape Velocity 0.163 km/s
Surface Temperature 64 K / –209°C
Mean Density 1.14 g/cm³

Ireland

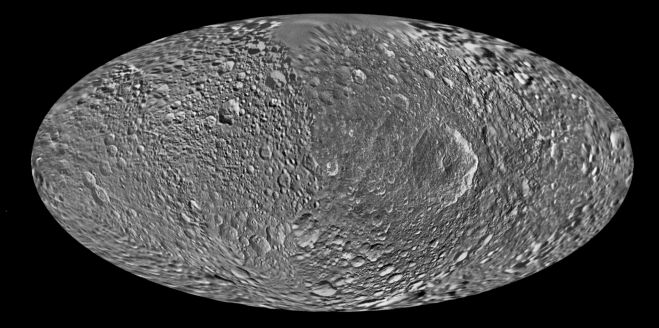

O lucky moon

THE ENERGY NEEDED to disintegrate a body completely is called its binding energy. This is defined as the energy required to drive all its constituents infinitely far apart. To shatter the Earth, for instance, would require the energy liberated by about a million billion of the largest H-bombs ever detonated.

An impacting body carries energy simply by virtue of its motion (if anyone has ever run into you head-on, you will appreciate this). To shatter a body entirely, the impactor's energy of motion must exceed the binding energy. The Moon is believed to have been created when a Mars-mass body hit the Earth. Since that body did not destroy our planet, it must have been moving unusually slowly, which has led to the suggestion that it actually shared the Earth's orbit.

Of course, whether the target body is shattered also depends on how fragile it is. An impactor that can shatter an iron moon needs to be moving only a fifth as fast to disintegrate an ice moon.

The size of a crater depends on the energy of motion of the impactor. You would think that a body capable of making a crater a third the size of Mimas would have shattered the moon. However, Mimas may have escaped such a fate because it is made of material that is very good at absorbing and dissipating energy. It is a lucky moon.

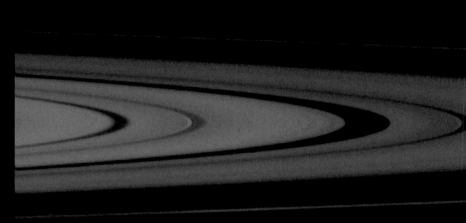

◀ The surface of Mimas shows a very strange temperature pattern, with the poles and one hemisphere warmer than the equatorial region of the other hemisphere. These regions must have surface compositions with different thermal capacity.

▶ This detailed color image mosaic shows subtle color variations around the crater Herschel on Mimas.

▲ Mimas is dwarfed by its parent planet, its blue winter hemisphere festooned with ring shadows in this view from Cassini.

◄ Mimas is seen here with the dark side of Saturn's rings and the irregular moon Epimetheus in the background.

► The moons Phobos, Mimas, and Tethys all sport large craters in relation to their size. Mimas's crater Herschel is 130 kilometers wide — more than one quarter the size of the moon itself.

▲ Phobos (26km) with its large crater Stickney (9km)

▲ Mimas (398km) and its crater Herschel (130km)

▲ Thethys (1,006km) and its impact basin Odysseus (400km)

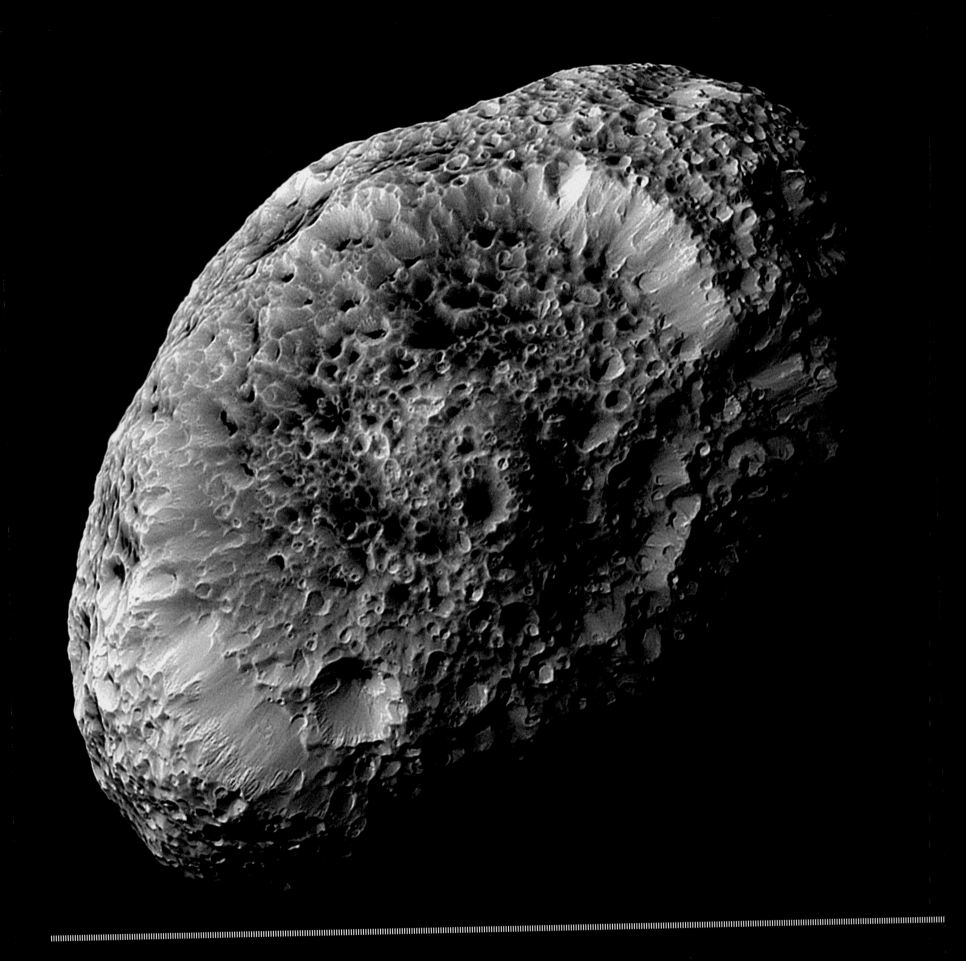

Hyperion

IT LOOKS EXACTLY like a piece of volcanic pumice stone. It would indeed be perfect for rubbing off a spot of hard skin if it were not 300 kilometers or so across. Hyperion is one of the most extraordinary and beautiful rocky bodies in the Solar System. The moon's low density indicates that its interior of rock and ice is filled with cavities. It may be a fragment of a larger body shattered by an impact. But its appearance is not Hyperion's only unusual feature. Weirder by far is the way it flies through space.

▼ Hyperion airbrush map based on images from the Voyager 2 mission. (Mollweide projection map centered on longitude 90° West.)

ORBITAL DATA

Distance from Saturn 1,330,000 to 1,640,000 km
Orbital Period (Year) 21.28 Earth days
Length of Day chaotic
Orbital Speed 5.6 to 4.6 km/s
Orbital Eccentricity 0.1042
Orbital Incination 0.43°
Axial Tilt chaotic

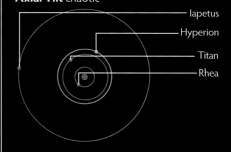

- Iapetus
- Hyperion
- Titan
- Rhea

PHYSICAL DATA

Diameter 266 km / 0.021 x Earth
Mass 5,580 million million metric tons
Gravity 0.002 x Earth
Escape Velocity 0.075 km/s
Surface Temperature 93 K / –180 C
Mean Density 5.666 g/cm³

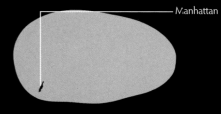

- Manhattan

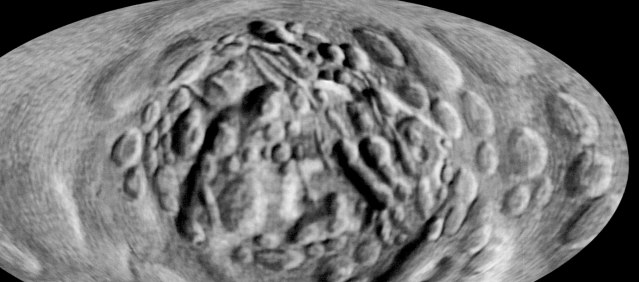

Chaotic moon

THE EARTH SPINS on its axis once every 24 hours, an axis that, just like a spinning top's, keeps pointing doggedly in the same direction. But imagine that the Earth were suddenly to slow down and stop spinning altogether, then start spinning once every 10 days, speeding up, slowing down again, all completely unpredictably. And, at the same time, pointing its spin axis first in one direction, then another, again unpredictably, so that the Earth twists and tumbles erratically through space.

Surely no astronomical body can behave so crazily? Well, actually, it can. Step forward Hyperion.

The major moons of the Solar System were born spinning at all manner of different rates. As time has passed, gravitational forces exerted by their parent planets have forced them to orbit, like our own Moon, with one face permanently toward their master.

Hyperion is prevented from reaching this tidally locked state by two things: its highly irregular shape–it is almost twice as long as it is wide; and the gravity of the giant moon Titan. These two factors conspire continually to change the forces on the moon in an unpredictable manner, preventing it from ever settling into a calm and stable state.

In 1984 Jack Wisdom and his colleagues predicted that Hyperion's motion would be inherently wild and unpredictable. They had found the first conclusive evidence of chaos in the Solar System.

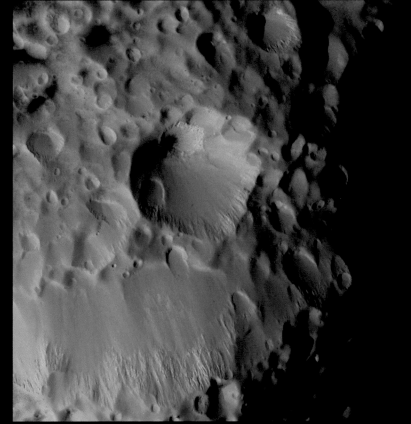

▼ Approaching Hyperion for the first time in 2005, the Cassini space probe took a sequence of images recording the moon's chaotic tumbling.

▶ Hyperion's crater Meri is the focus of this extreme color-enhanced view, revealing details down to 100 meters across.

 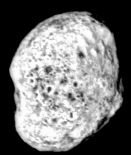 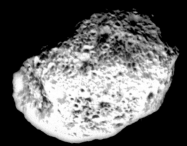 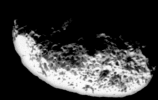

Tethys

Texas

The strangest moons

SATURN'S MOONS FALL roughly into three groups. One set congealed out of the debris disk around the still-forming planet, pretty much as planets congealed in the debris disk around the young Sun. These moons–including Titan, Rhea, Telesto, Dione, and Phoebe–are like a mini Solar System. And, just as in the Solar System, where Jupiter shares its orbit with Trojan asteroids that perpetually travel 60 degrees behind and ahead of the planet, Saturn's Tethys has two Trojan moons, Telesto and Calypso.

The orbits of a second group of Saturnian moons extend from within the inner edge of the rings to the outside edge. These pairs of shepherd moons, such as Pandora and Prometheus, strongly influence the structure of the rings, sculpting them with their gravity. They orbit, one moon on either side of a ringlet, shepherding the material into a narrow orbital path.

Perhaps the most interesting of such moons are Pan and Epimetheus, which share pretty much the same orbit. One circles Saturn about 50 kilometers inside the orbit of the other. Every four years or so, the inner moon gains on the outer one, overtaking it. When this happens, the outer moon tugs the inner one outward, and vice versa, and the two moons swap places. The whole dance, which is seen nowhere else in the Solar System, then repeats.

The third group of Saturnian moons orbit far from the planet. These are small and are likely to be cometary nuclei that have been captured by Saturn's gravity. Some orbit the opposite way to Saturn's spin, further bolstering the suspicion that they are interlopers, not born in a debris disk swirling around with the planet.

TOTAL NUMBER OF MOONS

62 (Tarqeq, **Pan**, Daphnis, **Atlas**, Prometheus, **Pandora**, Epimetheus, **Janus**, Aegaeon, **Mimas**, Methone, **Anthe**, Pallene, **Enceladus**, Tethys, **Calypso**, Telesto, **Polydeuces**, Dione, **Helene**, Rhea, **Titan**, Hyperion, **Iapetus**, Kiviuq, **Ijiraq**, Phoebe, **Paaliaq**, Skathi, **Albiorix**, S/2007 S2, **Bebhionn**, Erriapo, **Siarnaq**, Skoll, **Tarvos**, Greip, **S/2004 S13**, Hyrrokkin, **Mundilfari**, S/2006 S1, **Jarnsaxa**, Narvi, **Bergelmir**, S/2004 S17, **Suttungr**, Hati, **S/2004 S12**, Bestla, **Farbauti**, Thrymr, **S/2007 S3**, Aegir, **S/2004 S7**, S/2006 S3, **Kari**, Fenrir, **Surtur**, Ymir, **Loge**, S/2009 S1, and **Fornjot**)

Uranus

URANUS IS THE seventh planet from the Sun. It orbits 19 times as far away from the Sun as the Earth. Like Jupiter and Saturn, it is a gas giant–though it is slightly smaller in size than either and, of course, colder. Uranus was the first planet to be discovered that was entirely unknown to the ancients. Its discovery was a sensation.

ORBITAL DATA
Distance from Sun 2,750 to 3,000 million km / 13.85 to 20.02 AU
Orbital Period (Year) 84.01 Earth years
Length of Day 17.193 Earth hours
Orbital Speed 7.09x1000 to 6.51x1000 km/s
Orbital Eccentricity 0.0429
Orbital Incination 0.77°
Axial Tilt 97.92°

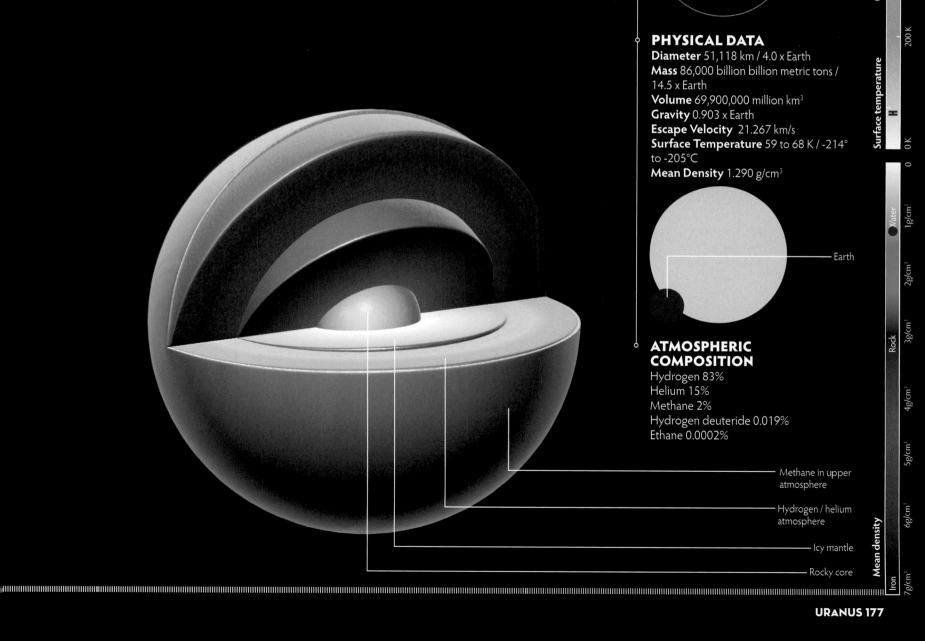

— Neptune
— Uranus
— Saturn
— Jupiter

PHYSICAL DATA
Diameter 51,118 km / 4.0 x Earth
Mass 86,000 billion billion metric tons / 14.5 x Earth
Volume 69,900,000 million km³
Gravity 0.903 x Earth
Escape Velocity 21.267 km/s
Surface Temperature 59 to 68 K / -214° to -205°C
Mean Density 1.290 g/cm³

— Earth

ATMOSPHERIC COMPOSITION
Hydrogen 83%
Helium 15%
Methane 2%
Hydrogen deuteride 0.019%
Ethane 0.0002%

Methane in upper atmosphere

Hydrogen / helium atmosphere

Icy mantle

Rocky core

Surface temperature

800 K
400°C
600 K
200°C
400 K
100°C
0°C
200 K
0 K

Mean density

Water
0
1g/cm³
2g/cm³
Rock
3g/cm³
4g/cm³
5g/cm³
6g/cm³
Iron
7g/cm³

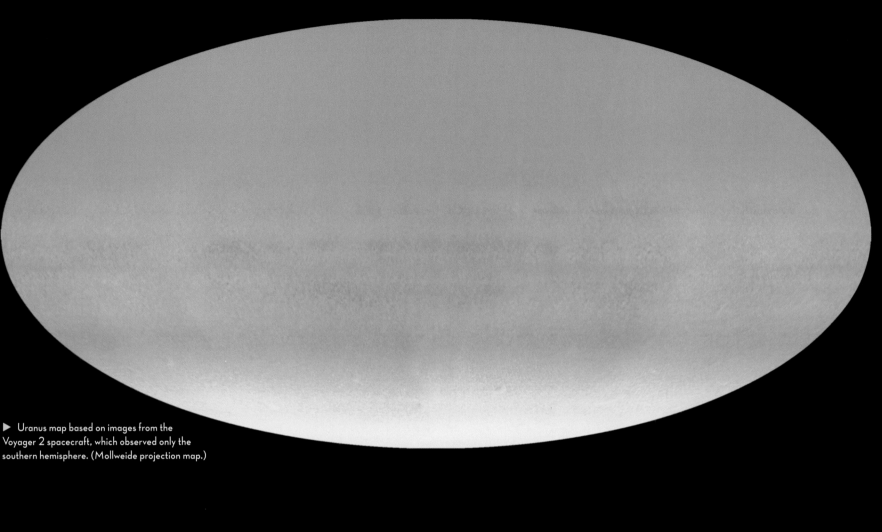

▶ Uranus map based on images from the Voyager 2 spacecraft, which observed only the southern hemisphere. (Mollweide projection map.)

▲ Uranus appears quite featureless at visible wavelengths. The blue-green color is due to absorption of red light by methane.

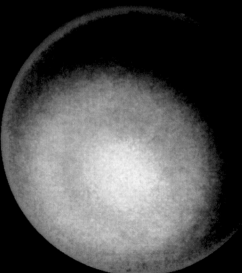

▲ Applying a strong contrast enhancement reveals some brightness variations in the upper atmosphere of Uranus, with a slightly brighter band around the south pole.

▲ Using ultraviolet as well as visible wavelengths reveals color varying from the pole to the Equator of Uranus.

▲ Boosting the brightness and contrast of this image from 2005 reveals the faint outer rings of Uranus.

▼ Around the time of Uranus's equinox, in 2007, the planet's faint rings appeared edge-on. A long exposure was required to record the full inner and outer ring system.

▼ The long exposure taken by Voyager 2 in 1986 shows the nine rings of Uranus known at that time.

▼ Voyager 2 discovered additional rings and dust lanes as it passed through the Uranian system.

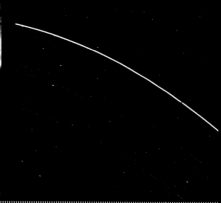

A planet called George

WILLIAM HERSCHEL was a freelance German musician. In 1757, at the age of 19, he moved to Bath, a town in England founded by the Romans because of its hot springs.

Although Herschel worked as an organist, he was fascinated by astronomy. His telescopes were the best in the world at the time. And it was on March 13, 1781, while in his garden scanning the night sky with his latest instrument, that a fuzzy star popped into his eyepiece. At first Herschel thought it was a comet. But over successive nights as it crept across the constellation of Gemini, Herschel realized it was following not the highly elongated orbit of a comet but the near–circular orbit of a planet.

Herschel named the planet George

(George's star), after the monarch of his adopted country, King George III. The French objected and referred to it as Herschel. The dispute was finally settled by the German astronomer Johann Bode, who suggested Uranus, father of the Roman god Saturn.

The discovery of a new planet created an international sensation. Uranus was twice as far from the Sun as Saturn, the most distant known planet. Overnight Herschel had doubled the size of the Solar System. Tantalizingly, the English astronomer John Flamsteed had catalogued Uranus almost a century earlier–but mistakenly, as a star.

It took telescopes much better than Herschel's to discover Uranus's peculiarity.

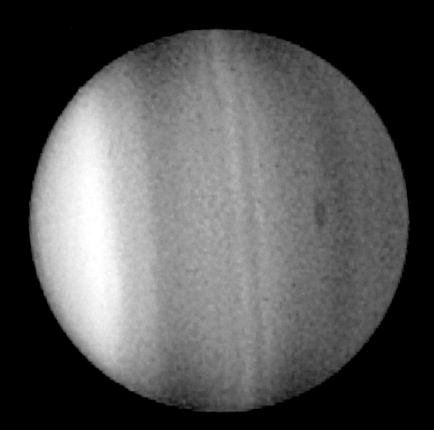

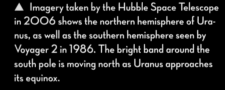

▲ Imagery taken by the Hubble Space Telescope in 2006 shows the northern hemisphere of Uranus, as well as the southern hemisphere seen by Voyager 2 in 1986. The bright band around the south pole is moving north as Uranus approaches its equinox.

▲ Although an organist, William Herschel still found time to discover Uranus.

Why is it on its side?

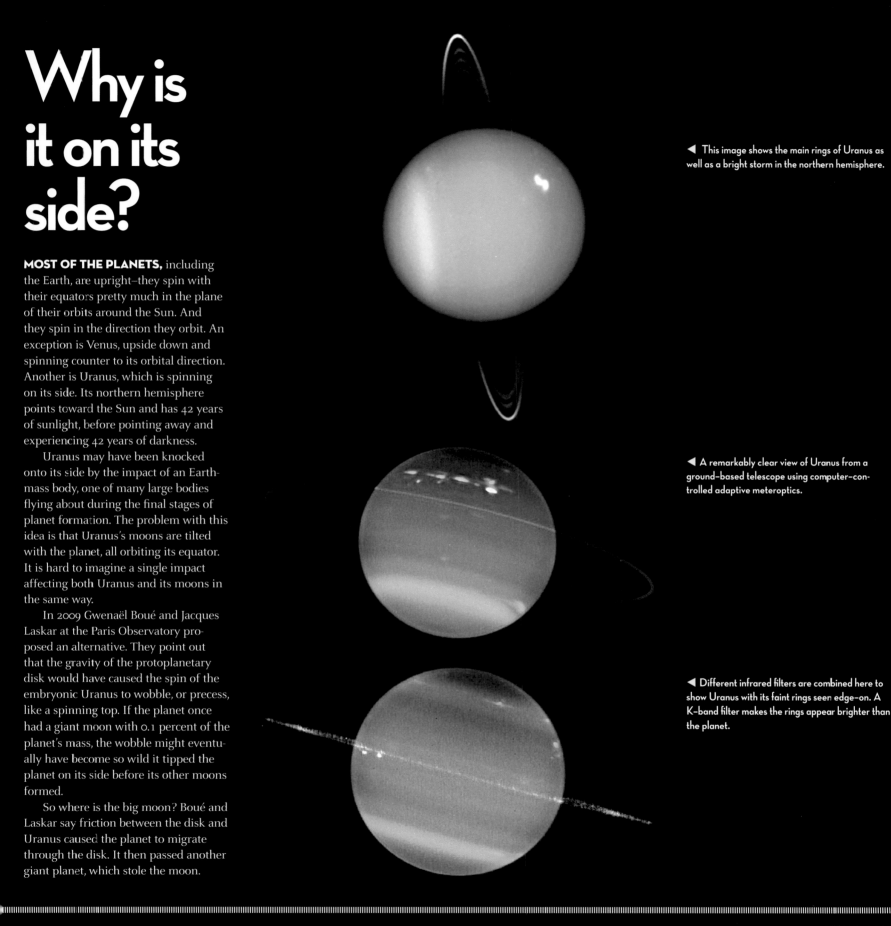

MOST OF THE PLANETS, including the Earth, are upright–they spin with their equators pretty much in the plane of their orbits around the Sun. And they spin in the direction they orbit. An exception is Venus, upside down and spinning counter to its orbital direction. Another is Uranus, which is spinning on its side. Its northern hemisphere points toward the Sun and has 42 years of sunlight, before pointing away and experiencing 42 years of darkness.

Uranus may have been knocked onto its side by the impact of an Earth-mass body, one of many large bodies flying about during the final stages of planet formation. The problem with this idea is that Uranus's moons are tilted with the planet, all orbiting its equator. It is hard to imagine a single impact affecting both Uranus and its moons in the same way.

In 2009 Gwenaël Boué and Jacques Laskar at the Paris Observatory proposed an alternative. They point out that the gravity of the protoplanetary disk would have caused the spin of the embryonic Uranus to wobble, or precess, like a spinning top. If the planet once had a giant moon with 0.1 percent of the planet's mass, the wobble might eventually have become so wild it tipped the planet on its side before its other moons formed.

So where is the big moon? Boué and Laskar say friction between the disk and Uranus caused the planet to migrate through the disk. It then passed another giant planet, which stole the moon.

◄ This image shows the main rings of Uranus as well as a bright storm in the northern hemisphere.

◄ A remarkably clear view of Uranus from a ground-based telescope using computer-controlled adaptive meteoptics.

◄ Different infrared filters are combined here to show Uranus with its faint rings seen edge-on. A K-band filter makes the rings appear brighter than the planet.

Miranda

MIRANDA IS A small icy moon a mere 470 kilometers across. Known as the patchwork moon, its surface is quite extraordinary.

ORBITAL DATA
Distance from Uranus 129,000 to 130,000 km
Orbital Period (Year) 1.41 Earth days
Length of Day 1.41 Earth days
Orbital Speed 7.1 to 6.5 km/s
Orbital Eccentricity 0.0013
Orbital Incination 4.23°
Axial Tilt 0°

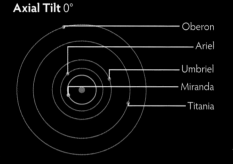

- Oberon
- Ariel
- Umbriel
- Miranda
- Titania

PHYSICAL DATA
Diameter 472 km / 0.04 x Earth
Mass 65,900 million million metric tons
Volume 55 million km^3
Gravity 0.008 x Earth
Escape Velocity 0.193 km/s
Surface Temperature 50 to 86 K / -223° to -187°C
Mean Density 1.150 g/cm^3

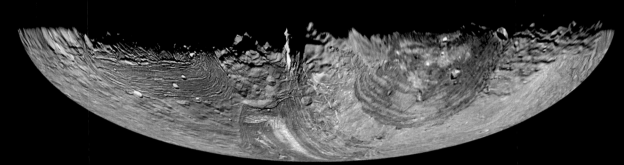

▲ Miranda map based on images from the Voyager 2 spacecraft, which observed only the southern hemisphere. (Mollweide projection map centered on 0° longitude.)

Ireland

Surface temperature

800 K
400°C
600 K
200°C
400 K
100°C
200 K
0°C
0 K

Mean density

0
1g/cm^3
2g/cm^3
3g/cm^3
4g/cm^3
5g/cm^3
6g/cm^3
7g/cm^3

Rock

Iron

Patchwork moon

NOBODY HAD SEEN a moon like it. Its icy surface was a crazy jumble of different terrains. It looked for all the world as if someone had gotten a hammer, smashed the moon apart, then glued the fragments back together—while blindfolded.

The irony is that Miranda was not even on the priority list to be imaged. But NASA's Voyager 2 space probe, in order to get to Neptune, had to boost its velocity by swinging in very close to Uranus. As it catapulted back out into interplanetary space, its trajectory just happened to take it past Miranda, one of the planet's innermost moons. It was a piece of amazing good fortune.

Not perhaps for the planetary scientists on live TV who had to explain what the viewers were seeing as the images came in.

They were completely flummoxed.

Initially, they trusted their first impression. In a cataclysmic collision in the past, they suggested, Miranda had indeed been blown apart, and the pieces had then come back together again, reconstituting the moon. But this is an extremely unlikely scenario, requiring an impact violent enough to shatter a moon but soft enough that the pieces move slowly enough to be re-snared by one another's gravity.

Nowadays, planetary physicists favor another scenario. They think that heating, perhaps from tidal stretching caused by Uranus's gravity, has given the moon convulsions. Time after time, liquid mixed with fragments of ice has welled up onto the surface. But the mystery of Miranda is far from solved.

Uranus's moons

URANUS HAS 27 known moons. While other bodies in the Solar System derive their names from classical mythology, Uranus's moons are named after characters from the writings of William Shakespeare and Alexander Pope. The moons Oberon and Puck, for instance, are named after fairies in Shakespeare's play *A Midsummer Night's Dream,* whereas the satellites Ariel and Umbriel honor characters in Pope's poem *The Rape of the Lock.*

TOTAL NUMBER OF MOONS
27 (Cordelia, **Ophelia**, Bianca, **Cressida**, Desdemona, **Juliet**, Portia, **Rosalind**, Cupid, **Belinda**, Perdita, **Puck**, Mab, **Miranda**, **Ariel**, Umbriel, **Titania**, Oberon, **Francisco**, Caliban, **Stephano**, Trinculo, **Sycorax**, Margaret, **Prospero**, Setebos, and **Ferdinand**)

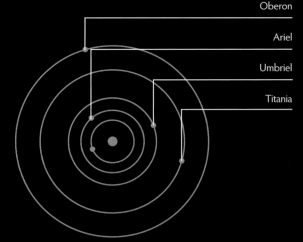

Oberon
Ariel
Umbriel
Titania

◀ Infrared image of Uranus and seven of its moons. At this wavelength, the moons and rings appear brighter than Uranus itself.

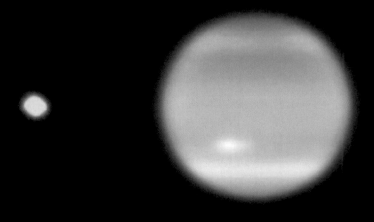

▼ Uranus with four of its largest moons (from right to left) Titania, Ariel, Miranda, and Umbriel.

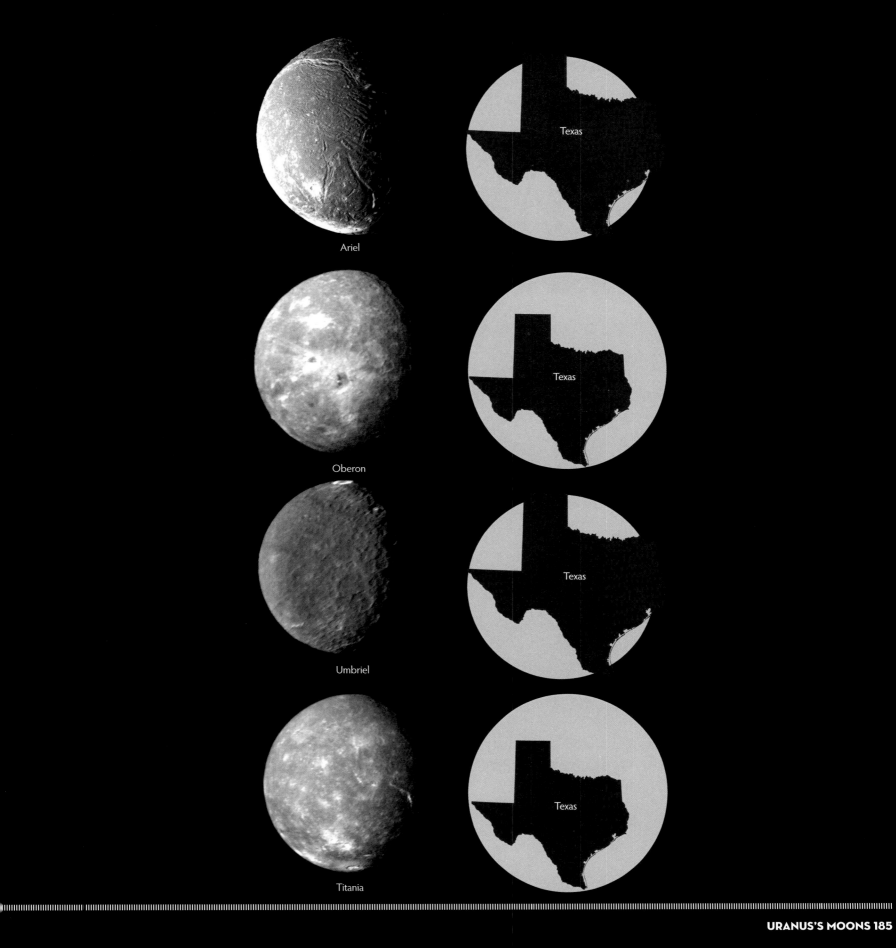

Ariel

Texas

Oberon

Texas

Umbriel

Texas

Titania

Texas

◀ This color image shows Ariel's southern hemisphere. The rims of the freshest craters show as bright spots.

◀ The best color view of Miranda, acquired by Voyager 2 in 1986.

▲ Ariel casts its shadow across the face of Uranus in this infrared view from the Hubble Space Telescope.

Mystery of the missing moon

URANUS'S MOONS, like Saturn's, fall roughly into three groups. The inner thirteen are associated with the planet's rings. They are tiny and were discovered by Voyager 2, the only space probe to have visited the planet, when it flew by on January 24, 1986.

The inner moons are made of the same dark, dusty material as the rings. This hints they are fragments of a larger moon that wandered so close to Uranus that it was ripped apart by the planet's gravity, spawning the ring system. Certainly, the moon Mab today appears to be re-supplying one of the rings with new dusty material.

By contrast, Uranus's outer nine moons are very likely bodies that wandered too close to Uranus and were captured into orbit by the planet's gravity.

In between Uranus's inner and outer moons are the planet's largest moons. These are believed to have formed out of the debris disk swirling around the newborn planet, just as planets congealed in the debris disk around the newborn Sun. The biggest of the five, Titania, is less than half the size of the Earth's moon. This poses a mystery since computer simulations of the birth of Uranus suggest it should have more substantial moons. Why does it not have at least one giant moon like its fellow gas giant planets Jupiter, Saturn, and Neptune?

The mystery may be solved if Uranus did start out with a giant moon but subsequently lost it to another planet. In fact, such a lost moon has recently been invoked by theorists to explain another puzzle: why Uranus is spinning on its side.

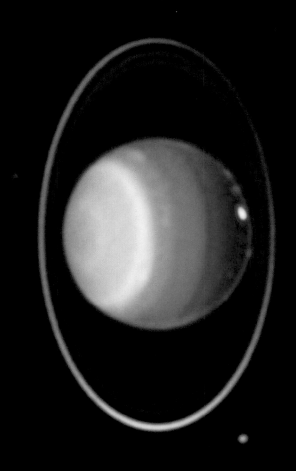

▲ Infrared imagery highlights several bright clouds in the northern hemisphere of Uranus. Also shown are the four main rings, and ten of Uranus' 17 moons.

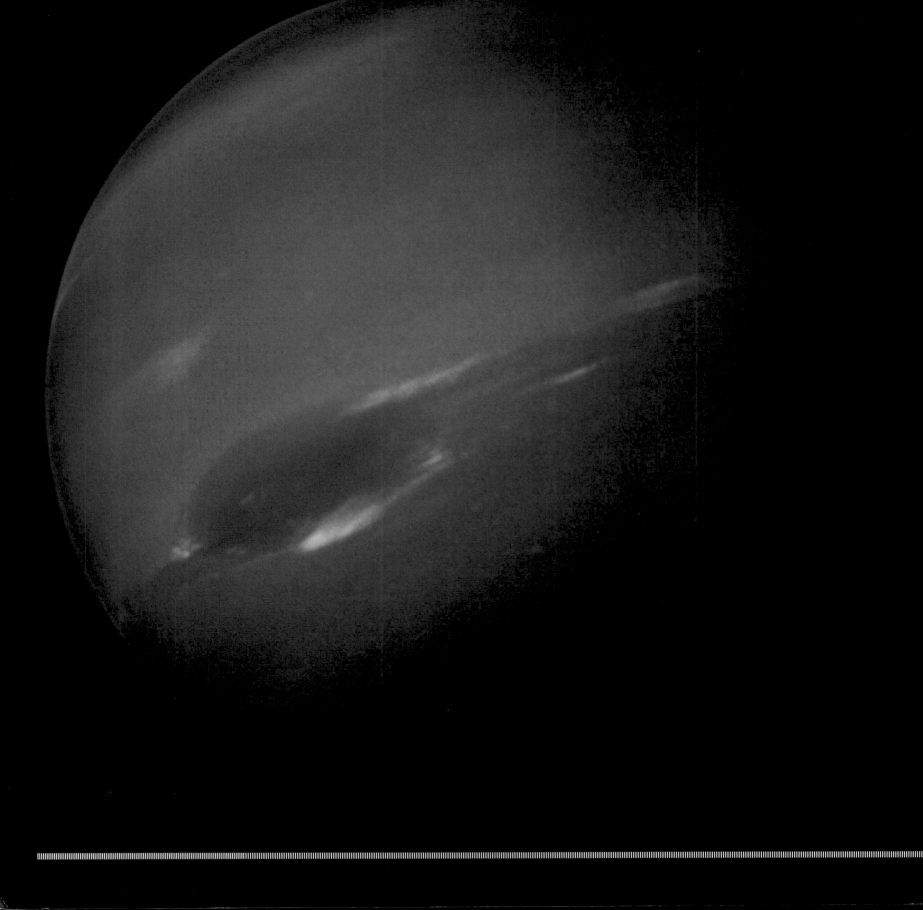

Neptune

NEPTUNE IS THE eighth planet from the Sun. Now that Pluto has been demoted, it is the outermost of the Solar System's planets.

Like Uranus, Neptune was entirely unknown to the ancients. It had to await the era of the telescope. In a triumph of Newtonian science, the Solar System's second "blue planet" was predicted before it was discovered.

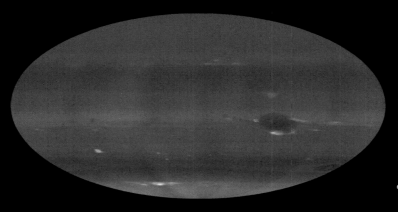

▲ Neptune map based on images from the Voyager 2 spacecraft. (Mollweide projection map.)

ORBITAL DATA

Distance from Sun 4,450 to 4,540 million km / 29.75 to 30.35 AU
Orbital Period (Year) 165 Earth years
Length of Day 16.1 Earth hours
Orbital Speed 5.5 to 5.4 km/s
Orbital Eccentricity 0.01
Orbital Incination 1.77°
Axial Tilt 28.8°

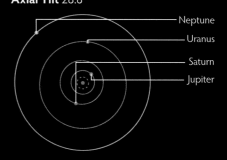

PHYSICAL DATA

Diameter 49,528 km / 3.88 x Earth
Mass 102,000 billion billion metric tons / 17 x Earth
Volume 625,000,000 million km^3 / 579 x Earth
Mean Density 1.640g/cm^3
Gravity 1.137 x Earth
Escape Velocity 23.491 km/s
Surface Temperature 50 to 53 K / -223° to -220° C

Earth

ATMOSPHERIC COMPOSITION

Hydrogen 80%
Helium 19%
Methane 1.5%
Hydrogen deuteride 0.019%
Ethane 0.00015%

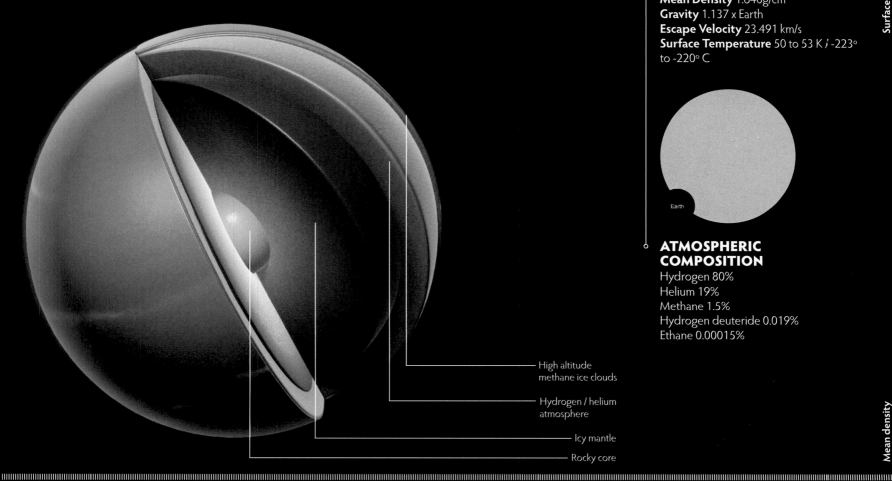

High altitude methane ice clouds

Hydrogen / helium atmosphere

Icy mantle

Rocky core

Dark matter of its day

MOST OF THE MATTER in the Universe is invisible, dark, matter. We know of its existence only because of the way its gravity tugs on the visible stars and galaxies. Neptune was the dark matter of its day.

By the early 19th century, astronomers were able to compare the observed orbit of Uranus with the orbit it should follow if solely under the influence of the sun. They found that Uranus's actual path deviated from expectation, the anomaly growing with time.

The suspicion arose that there must be another planet beyond Uranus tugging at it. In 1841 John Couch Adams, a young mathematical genius from Cornwall in England, set about the horrendous calculations necessary to deduce where in the sky the phantom planet must be. In 1845 he took his result to the Astronomer Royal, George Challis. But Challis did not take it seriously.

Meanwhile, in France, Urbain Le Verrier carried out similar calculations and took them to the director of the Paris Observatory, also without much effect. Impatient, Le Verrier sent his estimate of the rough location of the new planet to Johann Gottfried Galle in Berlin. On September 23, 1846, the German astronomer discovered Neptune.

Not surprisingly, there was a priority dispute between France and England–but when they eventually met, Adams and Le Verrier became firm friends. Nowadays, the discovery is attributed to them jointly.

The discovery of Neptune was a triumph for the theory of gravity. Not only did Newton's law explain what we could see, it also predicted what we could not see.

◀ Two Voyager images show Neptune's faint ring system without the glare from the much brighter planet.

Blue is the color

"PLANET EARTH IS blue and there's nothing I can do," sang David Bowie in "Space Oddity." The Earth from space looks blue or green-blue, because it is two-thirds covered in water. White light falling on the Earth from the Sun actually contains all the colors of the rainbow bundled together. Water absorbs them all, except blue-green, which it reflects. But what about other planets such as Neptune?

The color of a planet depends, in general, on what is doing the reflecting. In the case of a planet without an atmosphere–or even one with a moderate atmosphere, like the Earth–it is the planet's surface. In the case of a planet with an impenetrably thick atmosphere, it is the gases in the atmosphere.

Neptune is blue because of a small amount of methane in its atmosphere,

which absorbs red light from the Sun but reflects blue light back into space. Uranus has less methane and so appears less blue, more blue-green. Jupiter's orange bands are caused by ammonium sulfide, and its white bands by ammonia. Saturn's yellow comes from ammonia ice crystals in its atmosphere.

Clouds on Earth contain water droplets and ice crystals that are large enough to "scatter" all the colors of the rainbow, so they appear white, both from the ground and from space. Venus's impenetrable clouds are yellow, however, because this is the color reflected by the sulfuric acid of which they are made.

Mercury has no atmosphere and its surface reflects back a rocky gray. Mars is red because its surface–essentially rust–reflects the red in sunlight.

▼ The most detailed views we have of Neptune are from the Voyager 2 probe, which visited the planet in 1989.

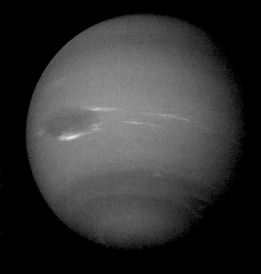

Storm World

THE GREAT THING about the Solar System is that it continually confounds our expectations. Solar heat drives the weather on Earth, so naively you might think that the farther from the Sun a planet, the calmer and duller its atmosphere. Nothing could be further from the truth. Neptune, the outermost planet, is the windiest place in the Solar System. The gales that scream around the blue planet's super-cold atmosphere blow at 2,000 kilometers an hour, six times faster than the fastest wind ever recorded on Earth.

Neptune's atmosphere is hyperactive. When NASA's Voyager 2 space probe flew by the planet on August 25, 1989, planetary scientists were amazed to see a Great Dark Spot–a raging storm in the southern hemisphere reminiscent of Jupiter's Great Red Spot–and a small irregular white cloud, dubbed the Scooter because it scooted so quickly around the planet.

By the time the Hubble Space Telescope observed the planet five years later, the Great Dark Spot had vanished. However, a new dark spot had appeared in the northern hemisphere.

On Jupiter, the energy to drive the weather comes from gravitational energy converted into heat energy as its interior slowly contracts. Neptune's interior is not contracting, so it poses more of a puzzle. One possibility is that gravitational energy is converted into heat as heavier liquid sinks to the center of the planet. This is just like the layers of oil and vinegar in salad dressing settling down after it is shaken. The truth? The blue planet's power source is yet another unexplained mystery of the Solar System.

◀ Long bright clouds stretch out along lines of latitude high above the blue of Neptune's main cloud deck.

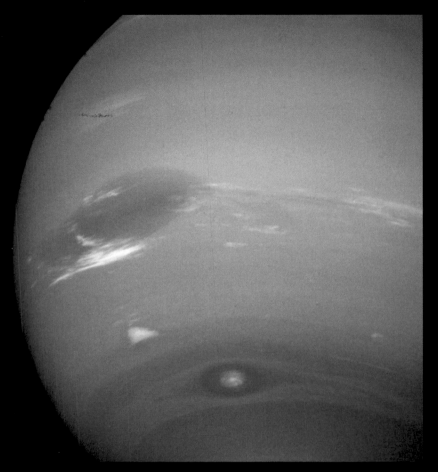

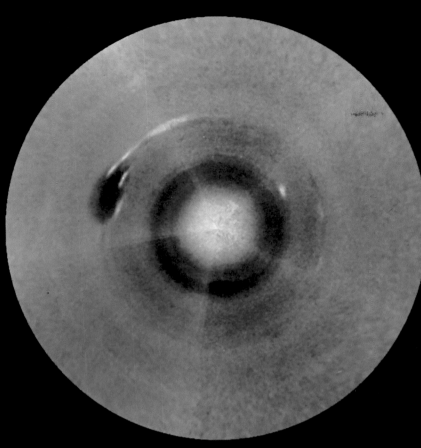

▲ A group shot of Neptune's main atmospheric features at the time of the Voyager encounter: The Great Dark Spot and its accompanying bright clouds; the Dark Spot 2 with its bright core; and between them a fast-moving bright feature named the "Scooter."

▲ A polar projection map of Neptune's southern hemisphere shows a bright pole surrounded by a dark band. The Great Dark Spot lies at 25 degrees south, about 28,000 kilometers from the pole.

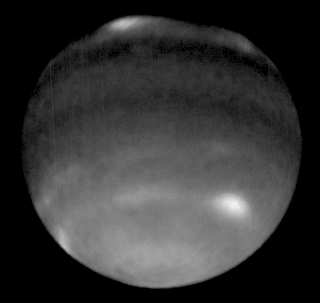

◀ Multiple wavelengths are combined to probe Neptune's atmosphere. High clouds show as white; even higher clouds are yellow–red.

▶ Triton map based on images from the Voyager 2 spacecraft, which observed only the southern hemisphere. (Mollweide projection map centered on 0° longitude.)

Triton

TRITON IS A PECULIAR and puzzling moon. On August 25, 1989, when NASA's Voyager 2 flew by the giant ball of ice and rock about two-thirds the size of our Moon, its images revealed geysers spewing matter into space. With the more recent discovery of ice fountains on Saturn's tiny Enceladus, Triton's geysers have become somewhat overshadowed. But at the time, they were a sensation. Unlike those on Enceladus and Io, they are not believed to be powered by tidal heating but merely by heat from sunlight concentrated under the solid nitrogen of the moon's polar cap. When the nitrogen turns to gas, it punches through cracks in the ice, rocketing to eight kilometers above the surface before being blown horizontally by thin winds. But the most extraordinary thing about Triton is not its geysers. It is its origin.

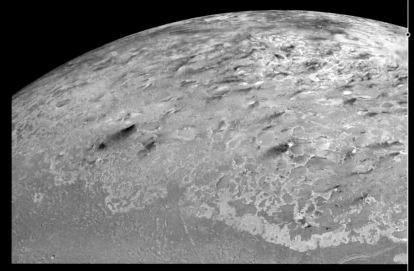

▶ Triton's south polar region is peppered by huge geyser-like plumes depositing dark, possibly carbonaceous, dust over the pink icy landscape.

Three–way pileup

IT CAME OUT of nowhere like an express train out of the night–not one world but a pair of worlds, caught in gravity's embrace. In the encounter with Neptune, one world was spun off to infinity, while the other was trapped forever.

Is this how Neptune acquired Triton? If so, it could explain the moon's peculiar orbit. Alone among the big planetary satellites, it orbits the wrong way–that is, Neptune spins one way on its axis, while Triton circles the planet the opposite way.

This was odd because big moons were thought to form out of debris disks swirling around a newborn planet, so they should swirl around in the same direction as the planet's rotation. Only small, "captured" moons can orbit in an opposite, "retrograde," manner, since they can be snared from any direction. Triton, however, is a big moon that has been captured.

There is an obvious source: the newborn Neptune would have had ample opportunity to encounter Kuiper Belt objects (KBOs). The trouble is that a body as massive as Triton could be captured only if it was moving extremely slowly, which is improbable.

But according to physicists Craig Agnor and Douglas Hamilton, there is a way around this if Triton had a companion. Their simulations show that, in the three–body encounter with Neptune, Triton could have lost speed at the expense of its companion, which would then have been ejected.

Many KBOs are known to be binaries. And it is highly suggestive that Pluto is about the size of Triton and its orbit crosses Neptune's. Could it be that Pluto and Triton are brothers?

ORBITAL DATA
Distance from Neptune 355,000 km
Orbital Period (Year) 5.88 Earth days
Length of Day 5.878 Earth days
Orbital Speed 4.4 km/s
Orbital Eccentricity 0
Orbital Incination 156.89°
Axial Tilt 0°

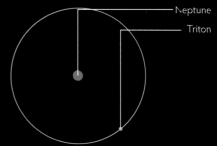

Neptune

Triton

PHYSICAL DATA
Diameter 2,707 km / 0.21 x Earth
Mass 21 billion billion metric tons / 0.004 x Earth
Volume 10,400 million km³ / 0.01 x Earth
Gravity 0.08 x Earth
Escape Velocity 1.453 km/s
Surface Temperature 38 K / –235°C
Mean Density 2.054 g/cm³

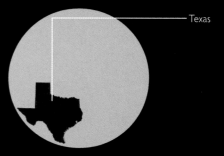

Texas

ATMOSPHERIC COMPOSITION
Nitrogen 99.999%
Carbon monoxide+Methane 0.001%

Surface temperature

800 K
600 K
400 K
200 K
0 K

400°C
200°C
100°C
0°C

Mean density

Water
Rock
Iron

0
1g/cm³
2g/cm³
3g/cm³
4g/cm³
5g/cm³
6g/cm³
7g/cm³

Kuiper Belt

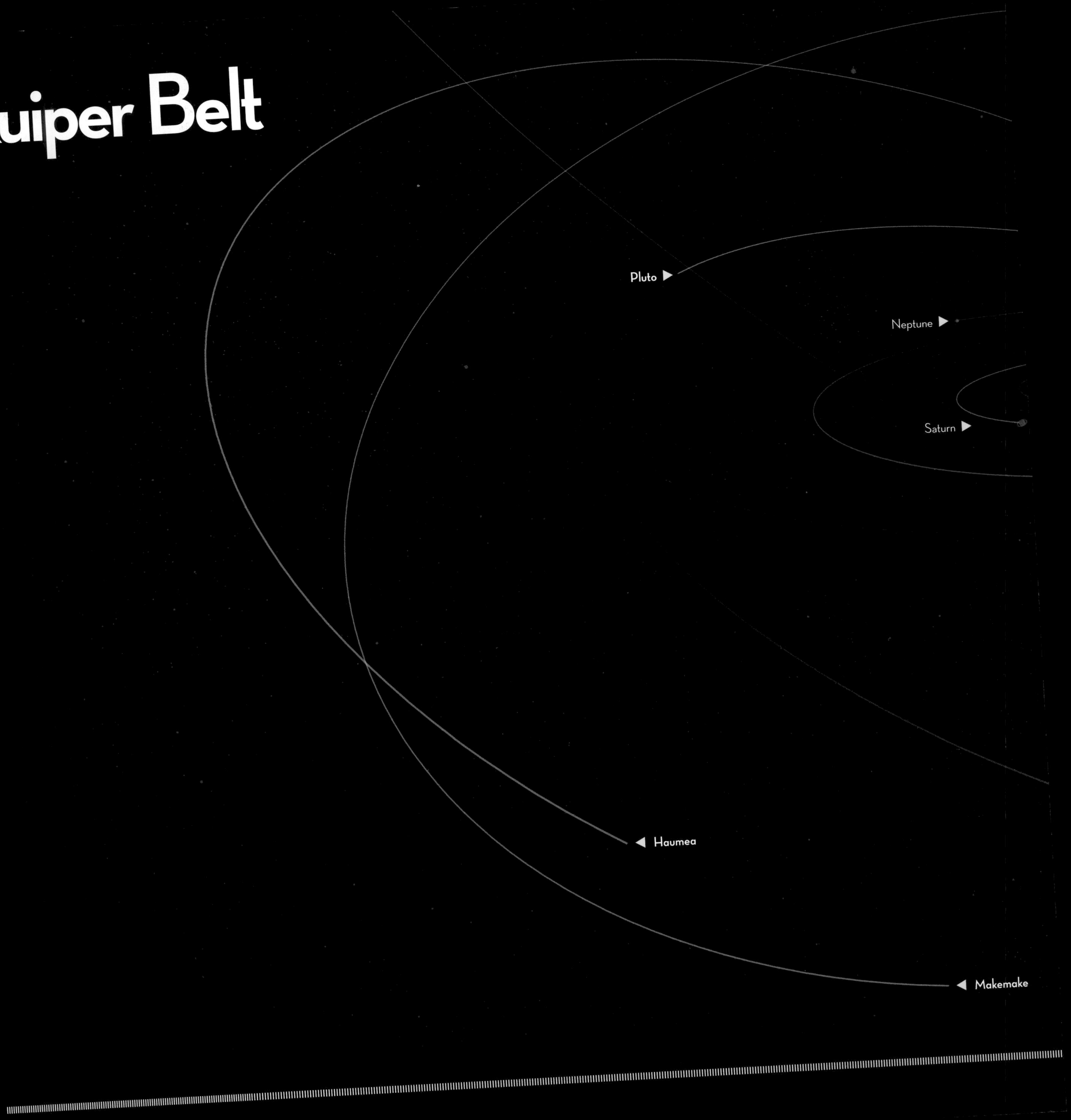

Pluto ▶

Neptune ▶

Saturn ▶

◀ Haumea

◀ Makemake

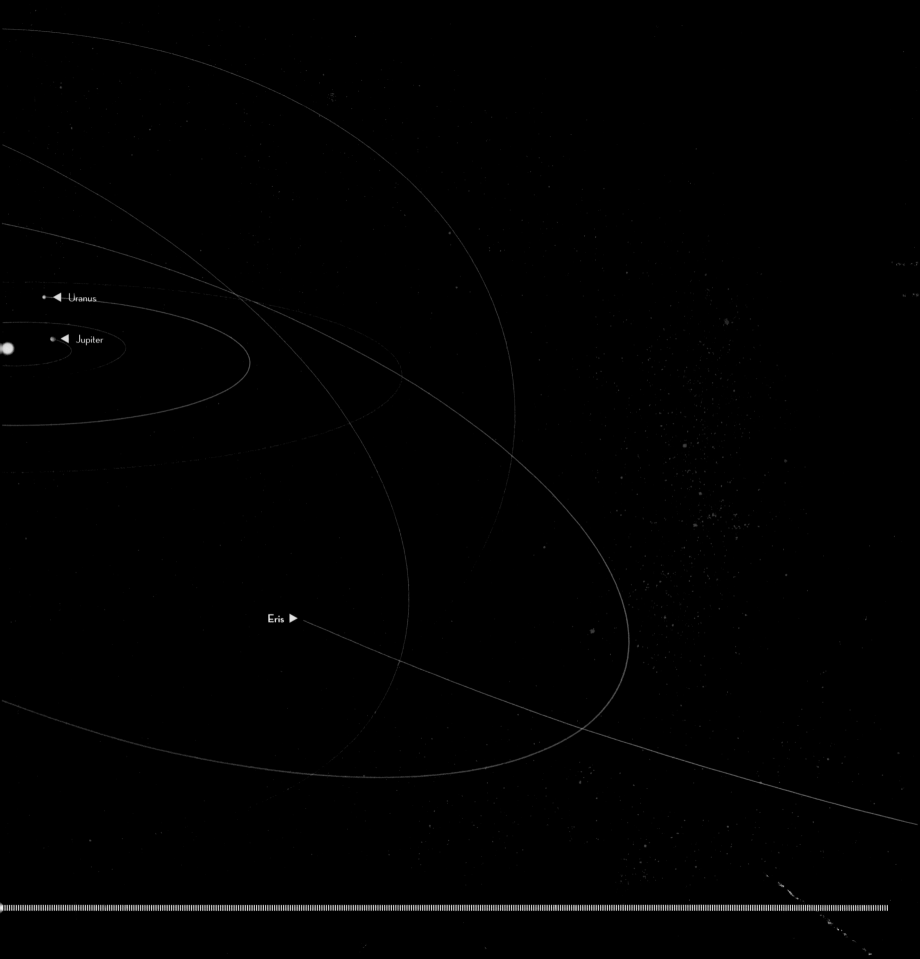

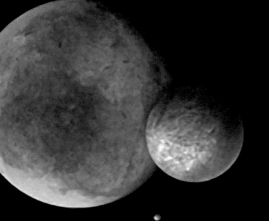
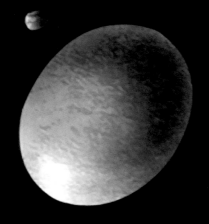
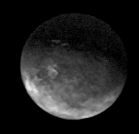

Kuiper belt

▲ The largest known Kuiper Belt objects, pictured to scale, arranged in order of increasing distance from the Sun (from left): Pluto, with its moons Charon, Nix, and Hydra; Haumea, with its moons Hi'iaka and Namaka; Quaoar; Makemake; Eris and its moon Gabrielle; and Sedna. Pluto, Eris, Makemake, and Haumea are classified as dwarf planets.

THE INNER SOLAR SYSTEM has the Asteroid Belt, rocky builders' rubble left over from the formation of the planets. The gravity of Jupiter prevented these rocks from aggregating into a proper world. The outer Solar System has the Kuiper Belt, icy builders' rubble left over from the birth of the planets, but in this case spread too thinly ever to form a planet. The inner edge of the Kuiper Belt, which is near Neptune, is about 30 times as far from the Sun as the Earth. The outer edge stretches as far as 50 times that distance.

So far, about 1,000 icy bodies have been discovered in the Kuiper Belt, among them objects hundreds or even thousands of kilometers across with names like Eris, Sedna, Makemake, Haumea, and Quaoar. The Kuiper Belt was actually predicted long before it was found.

NUMBER OF OBJECTS
70,000+ over 100km in diameter

LARGEST OBJECT
Pluto

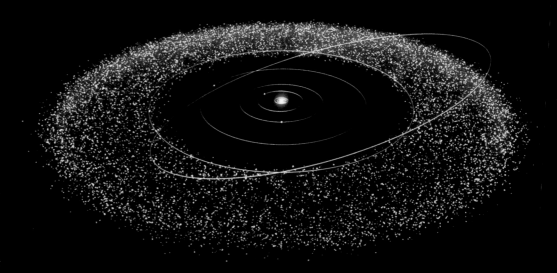

◄ Artist's impression of a Kuiper Belt object, orbiting the Sun beyond Neptune on the cold fringes of the Solar System.

▲ Simulation of the Kuiper Belt

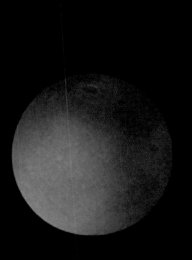

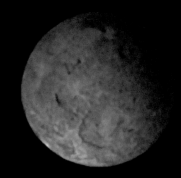

▲ Quaoar was discovered in 2002, orbiting the Sun at a distance of about 6 billion kilometers. Multiple images from the Hubble Space Telescope confirmed the presence of a planetoid moving against the background of stars.

▲ Sedna was discovered in 2004, orbiting 500 times further out from the Sun than does the Earth, making it the furthest known object in the solar system. It appears to move relative to the background star due to the motion of the space telescope in its orbit around the Earth, confirming that the object lies within the solar system.

Kuiper caper

IN 1943 a former Irish soldier and amateur astronomer was thinking about the birth of the Solar System. In the swirling debris disk around the newborn Sun lots of ice and rubble had collided and stuck together, building ever bigger bodies and eventually producing the planets. But it made no sense that the disk should have an edge, thought Kenneth Edgeworth. Surely it would have petered out slowly? Beyond Neptune, there should be a ring of leftover icy rubble, too sparse to have aggregated into planets.

The Dutch astronomer Gerrit Kuiper had a similar idea but never fleshed it out. Nevertheless, the icy ring gained Kuiper's name and not Edgeworth's, though some insist on calling it the Edgeworth-Kuiper Belt.

The Kuiper Belt solved a major comet puzzle. Short–period comets were believed to be long-period comets snared by Jupiter's gravity and trapped in the inner Solar System. Not only is such trapping too inefficient to explain their numbers but short-period comets, unlike long-period ones, orbit in the same plane as the planets. They therefore cannot come from the spherical Oort Cloud, the reservoir of long–period comets. There must be another source: a ring of orbiting icy debris in the outer Solar System–the Kuiper Belt.

A comet could be nudged out of the belt only by a close encounter with a big body. The Kuiper Belt must therefore contain large bodies too. At this point, a penny dropped for astronomers. A large icy body in the right location was already known: Pluto. Rather than a planet, it must be a Kuiper Belt object.

▲ Astronomer Gerrit Kuiper (1905-1973).

Pluto

▶ Pluto, as viewed by the New Horizons probe in July 2015.

PLUTO IS A SUPER-COLD ball of icy rock much smaller even than the Earth's Moon. Its discovery in 1930 hit newspaper headlines around the world. One of those newspapers, *The Times* of London, was being read by the grandfather of 11-year-old Venetia Burney as she munched her breakfast in Oxford, England. When he read aloud the report of the discovery of the ninth planet, Venetia thought for a while and then said: "They should call it Pluto." Pluto was the Roman god of the underworld. Venetia's grandfather was enthusiastic. That morning, he slipped a note through the letterbox of an Oxford astronomer, who passed on Venetia's suggestion to Vesto Slipher, director of the Lowell Observatory in Arizona, where the new planet had been found.

But before all this, Pluto had to be found.

▼ New Horizons shows Pluto's icy surface to be surprisingly rugged, with mountains rising up to 11,000 feet, and surprisingly crater-free, indicating a geologically youthful surface.

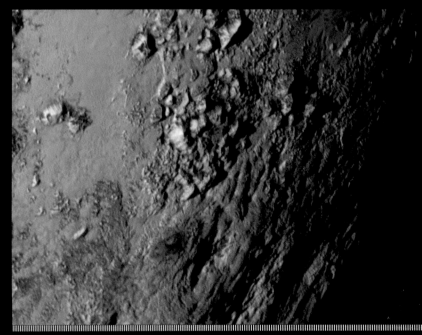

▲ Pluto is seen sometimes to be elongated in ground-based telescope views (left in this image). In 1978 this elongation was attributed to a companion moon, named Charon. A clearer picture is obtained by space telescopes operating high above the Earth's atmospheric aberrations (right).

◀ The telescope at Arizona's Lowell Observatory with which Clyde Tombaugh took photographs of the night sky at different times.

ORBITAL DATA

Distance from Sun 4,440 to 7,390 million km / 29.68 to 49.40 AU
Orbital Period (Year) 248.5 Earth years
Length of Day 6.38 Earth days
Orbital Speed 6.1 to 3.7 km/s
Orbital Eccentricity 0.25
Orbital Incination 17.12°
Axial Tilt 119.6°

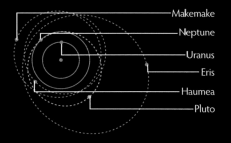

— Makemake
— Neptune
— Uranus
— Eris
— Haumea
— Pluto

PHYSICAL DATA

Diameter 2,304 km / 0.18 x Earth
Mass 13 billion billion metric tons / 0.002 x Earth
Volume 6,160 million km3 / 0.006 x Earth
Gravity 0.067 x Earth
Escape Velocity 1.227 km/s
Surface Temperature 33 to 55 K / -240° to 0218°C
Mean Density 2.050 g/cm³

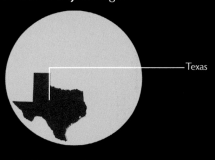

— Texas

ATMOSPHERIC COMPOSITION

Nitrogen 90%,
Carbon monoxide+Methane 10%

Out of the darkness

"YOUNG MAN, I am afraid you are wasting your time–if there were any more planets they would have been found long before this," an astronomer said to Clyde Tombaugh in 1929. Fortunately, Tombaugh took no notice. A farm boy from Kansas, he had written to the Lowell Observatory in Flagstaff, enclosing planet drawings he had made of Mars and Saturn. So impressed was the observatory's director, Vesto Slipher, that he had given the 23-year-old Tombaugh a job–a jaw-droppingly tedious job.

The eighth planet, Neptune, had been predicted because of an unexplained gravitational tug on Uranus. After many years of observation of Neptune, astronomers began to suspect–incorrectly–that its presence could not totally explain the anomaly of Uranus's orbit. Could there be a ninth planet, a so–called Planet X? Tombaugh was set the task of finding it.

The young astronomer painstakingly photographed the stars around the zodiac. He took two pictures of each region of sky, a few days apart, and flipped back and forth between them using a machine called a "blink comparator." Any object that "blinked" against the steady background of stars could be a nearby object.

Tombaugh's ten months of dedicated work would eventually yield 29,000 new galaxies, 3,969 asteroids, 1,800 variable stars, and 2 comets. On February 18, 1930 his amazing dedication paid off. There, winking back at him in the darkness, was a ninth planet. Tombaugh burst into the director's office. "Dr Slipher," he declared, "I have found your Planet X!" In time, it would be named Pluto.

▲ NASA's New Horizons probe reached Pluto after a nine-year journey in 2015, snapping this image of the dwarf planet with its moon Charon as it approached on July 8.

Surface temperature

800 K
400°C
600 K
200°C
400 K
100°C
200 K
0°C
0 K

Mean density

Iron 7g/cm³
6g/cm³
5g/cm³
4g/cm³
3g/cm³
2g/cm³
Water 1g/cm³
Rock
0

Dwarf planets

FROM THE MOMENT of Pluto's discovery there were doubts. It was far smaller than expected–too tiny to have any effect on the orbit of Uranus. Then there was the problem of its orbit, determined by looking back at old photographic plates that had inadvertently recorded it. Not only was Pluto's orbit steeply inclined to the plane in which all the other planets circled the Sun, it was also so elongated that it cut inside the orbit of Neptune. There were times when Pluto was not the ninth planet but the eighth.

All this was enough to ring alarm bells. But it was not until the 1990s, when it became widely accepted that there was a ring of icy debris orbiting in the outer Solar System, that astronomers began to think that Pluto could simply be an unusually large member of this Kuiper Belt.

The discovery of large numbers of Kuiper Belt objects, including Eris, which is possibly bigger than Pluto, meant it was time for action. They could not all be planets. On June 30, 2006, the International Astronomical Union's planet definition committee met at the Paris Observatory–in the office of Le Verrier, who had predicted Neptune. Controversially they demoted Pluto to the status of a dwarf planet. These are bodies orbiting the Sun massive enough to be spherical but not enough to have cleared their neighborhood of other rubble–and which are not moons.

Pluto joined four other dwarf planets: Ceres, Haumea, Makemake, and Eris. Fortunately, Tombaugh, who had died in 1997 aged 90, did not live to see the ignominious demotion of his beloved planet.

▲ Pluto's brown surface is thought to be due to deposits of frozen methane stained by exposure to sunlight. A coarse color map can be built up from images taken at different rotation angles, and as Charon eclipses Pluto.

▼From Pluto's surface, the Sun appears one thousand times fainter than it does from Earth.

◀ Pluto's largest moon Charon, imaged by the New Horizons probe from a distance of 289,000 miles. The image shows deep fractures and a dark polar cap.

▲ Pluto and Charon in false color, highlighting differences in surface composition.

▲ It is now known that Pluto has four smaller moons in addition to Charon: Hydra, Nix, Styx, and Kerberos.

Charon

IN RETROSPECT, it is amazing nobody noticed before: Pluto is not one planet but two. Photographic images taken over the years sometimes showed the tiny white blob slightly elongated in one direction or another, but astronomers put it down to bad atmospheric "seeing" or flaws in their photographic plates. Until June 22, 1978, that is.

Jim Christy, at the United States Naval Observatory in Flagstaff, Arizona, was looking at two photographic plates of Pluto under a microscope. The images were elongated, in different directions. But Christy noticed something odd: the background stars were perfectly sharp. The anomaly must therefore be an intrinsic property of Pluto. Suddenly, it hit him: Pluto had a moon.

The moon had to be huge compared with Pluto. And, from the time interval between the images, he could see it orbited in about six days.

Driving his car that night, Christy turned to his wife, Charlene, whom he called "Char," and jokingly said he would name Pluto's moon after her. He would call it Charon, which he pronounced "Sharon." She did not take him seriously. But that night Christy woke with a nagging thought. He grabbed his encyclopedia and, by torchlight, flicked through the pages. Charon was the ferryman who carried the dead across the River Styx to the underworld. It was perfect. He was convinced he had a winning name.

And so it proved.

The extraordinary thing about Charon is that it is almost as big as Pluto. This system is in effect a binary dwarf planet. Charon is not a moon orbiting a planet but a planet orbiting a planet.

▲ Slight color variations are apparent between Pluto and its moons. Charon is thought to be slightly bluer than Pluto, with more water ice on its surface.

Eris

ERIS IS A GIANT snowball of a world that plies the lonely, dark wastes on the frigid fringes of the Solar System. It is not alone. It is accompanied by a tiny moon called Dysnomia. Together the pair go around the Sun in a highly elongated orbit. At their closest they are 38 times farther from the Sun than the Earth and at their farthest 97 times. The orbit of Eris and Dysnomia is so large and their motion so sluggish that they take 557 years to complete one circuit of the Sun.

Eris, originally designated object 2003 UB 313, was discovered in January 2005, although the images that revealed it were taken in 2003. It is fair to say that, when astronomers Mike Brown, Chad Trujillo, and David Rabinowitz spotted the tiny speck of light crawling across the backdrop of stars, they never guessed the furor it would cause.

◄ This image from 2006 is sharp enough to show Eris and its moon Dysnomia. Eris was discovered in 2003, Dysnomia two years later.

Pluto killer

THE DISCOVERY OF object 2003 UB 313 in January 2005 was a bombshell dropped into the world of astronomy. Not only did it orbit far beyond Pluto but it soon became apparent that it might be slightly bigger than Pluto. Should it be declared the tenth planet?

The trouble was, object 2003 UB 313—briefly called Xena, after a TV warrior princess, before it was christened with its official title of Eris—was associated with the Kuiper Belt. It was not exactly in the Kuiper Belt—it was in a highly elongated orbit just outside—but it had clearly originated there. And the Kuiper Belt could contain several hundred bodies comparable in size to Pluto.

Could astronomy really tolerate a Solar System with several hundred planets?

The discovery of Eris was the catalyst for a major rethink of the constitution of the Solar System. It culminated in the June 30, 2006 meeting of the International Astronomical Union's planet definition committee in Paris in which Pluto was downgraded to the status of a dwarf planet. Eris joined it.

Pluto had survived as a planet in its own right for 76 years. Its discoverer, Clyde Tombaugh, died still believing he had found the ninth planet. But Pluto had finally succumbed to the inevitable. And Eris was its killer.

ORBITAL DATA
Distance from Sun 5,650 to 14,600 million km / 37.77 to 97.59 AU
Orbital Period (Year) 556 Earth years
Length of Day 8 Earth hours
Orbital Speed 5.8 to 2.3 km/s
Orbital Eccentricity 0.4418°
Orbital Incination 44.19°

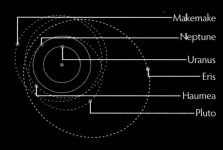

— Makemake
— Neptune
— Uranus
— Eris
— Haumea
— Pluto

PHYSICAL DATA
Diameter 2,600 km / 0.20 x Earth
Mass 17 billion billion metric tons / 0.003 x Earth
Volume 9,200 million km³ / 0.009 x Earth
Gravity 0.067 x Earth
Escape Velocity 1.309 km/s
Surface Temperature 27 to 43 K / -246° to -230°C
Mean Density 2.3 g/cm³

The Moon

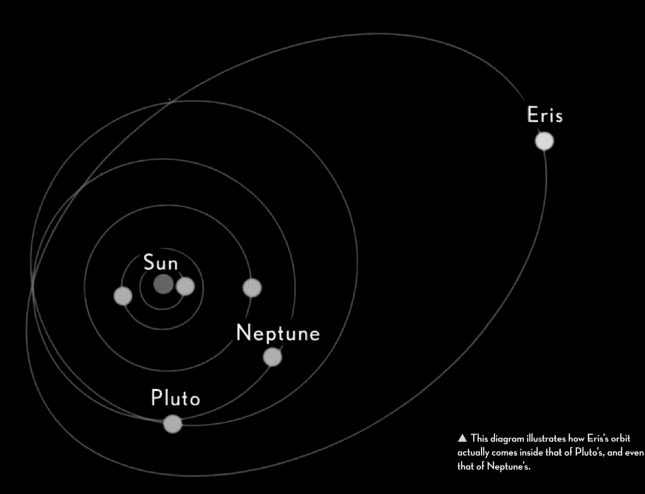

Eris

Sun

Neptune

Pluto

▲ This diagram illustrates how Eris's orbit actually comes inside that of Pluto's, and even that of Neptune's.

Surface temperature

800 K
400°C
600 K
200°C
400 K
100°C
0°C
200 K
0 K

Mean density

0
Water 1g/cm³
2g/cm³
Rock 3g/cm³
4g/cm³
5g/cm³
6g/cm³
Iron 7g/cm³

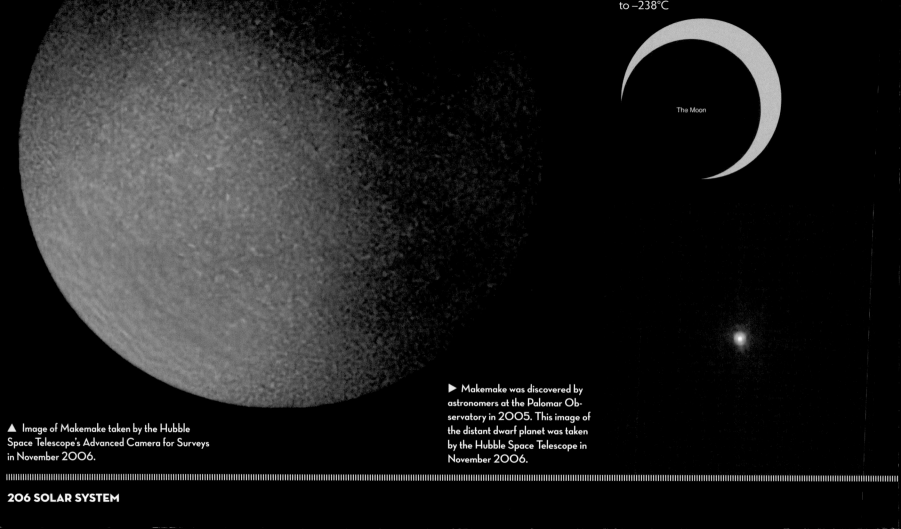

to −238°C

The Moon

▲ Image of Makemake taken by the Hubble Space Telescope's Advanced Camera for Surveys in November 2006.

▶ Makemake was discovered by astronomers at the Palomar Observatory in 2005. This image of the distant dwarf planet was taken by the Hubble Space Telescope in November 2006.

Haumea

HAUMEA, NAMED AFTER a Hawaiian goddess, was discovered on December 28, 2004. Although only a third of the mass of tiny Pluto, it was nevertheless designated a "dwarf planet" by the International Astronomical Union on September 17, 2008. What makes it unique among the four dwarf planets is its shape—it is twice as long as it is wide. This is believed to be not because it is a broken fragment of a bigger body like a potato-shaped asteroid but because it is spinning extremely rapidly, its icy surface bulging out around its equator. If it were not spinning so fast, it would be a sphere like all other dwarf planets.

Nobody knows why Haumea is spinng so fast but it could be the result of a collision with another body. Its two tiny moons, Hi'iaka and Namaka, might then be "chips off the old block," created in the collision.

▼ Image of Haumea and its two moons, Hi'iaka and Namaka, taken by the 10 meter-diameter Keck II telescope, using the Keck Observatory Laser Guide Star Adaptive Optics system, in 2005.

ORBITAL DATA
Distance from Sun 5,190 to 7,710 km / 34.69 to 51.54 AU
Orbital Period (Year) 283 Earth years
Length of Day 3.916 Earth hours
Orbital Speed 5.5 to 3.7 km/s
Orbital Eccentricity 0.195°
Orbital Incination 28.22°

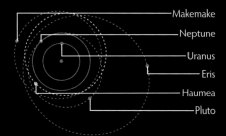

- Makemake
- Neptune
- Uranus
- Eris
- Haumea
- Pluto

PHYSICAL DATA
Diameter 1,436 km / 0.11 x Earth
Mass 4 billion billion metric tons / 0.001 x Earth
Gravity 0.053 x Earth
Escape Velocity 0.862 km/s

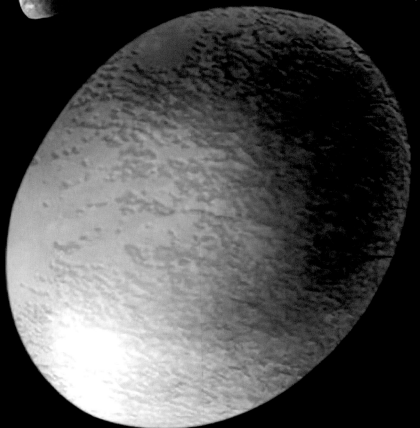

The Moon

▲ Haumea and its two moons are shown in this view from the W M Keck Observatory in Hawaii: Hi'iake is above, Namaka is just visible below.

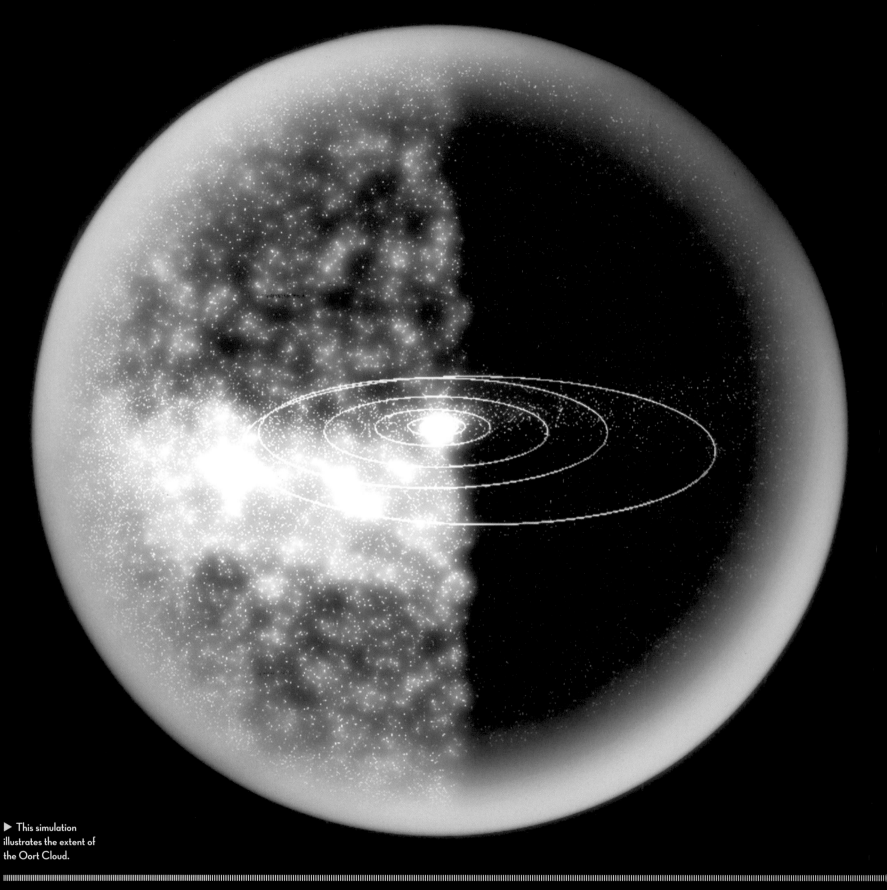

▶ This simulation illustrates the extent of the Oort Cloud.

Oort Cloud

UNLIKE ALL THE OTHER components of the Solar System, this one nobody has seen—nor is likely to see. Not until we venture out to the stars. Yet it utterly dwarfs the realm of the planets and contains the majority of the bodies in the Solar System. The Oort Cloud is a giant swarm of cometary nuclei.

We know it exists, even though we have not seen it. How?

DIAMETER
300,000 to 3,000,000 million km / 2,000 to 20,000 AU (inner Oort Cloud)

3 to 7.5 million million km / 20,000 to 50,000 AU (outer Oort Cloud)

OBJECTS > 1 km
Several trillion (estimated)

OBJECTS > 20 km
Many billion (estimated)

Why do we think it is there?

WHERE DO COMETS come from? The question nagged at Dutch astronomer Jan Oort shortly after the Second World War.

The clue came from "long-period" comets, which plunge into the inner Solar System, swing around the Sun, and fly back out, probably never to be seen again. When Oort studied their orbits, he discovered that, unlike those of planets, they were not confined to a thin disc around the Sun. Comets plunge in from above or below the plane of the planets from any direction at all.

The conclusion, to Oort, was unavoidable. Far beyond the planets there must be a gigantic reservoir of comets. Oort imagined it as a giant spherical swarm of icy bodies, stretching halfway to the nearest star. The planets were embedded in this Oort Cloud but totally dwarfed by it.

Oort knew how many comets came into the inner Solar System in a century and he knew they had been doing so for billions of years. He did a calculation. The cloud could contain anywhere from 10 billion to a trillion comets, each taking a million years or so to wheel around the Sun. Occasionally, the gravity of a passing star would nudge a comet, sending it plunging sunward.

But the gas cloud that spawned the Solar System had shrunk from a sphere to a flattened disk. How come the comets are distributed spherically? It must be that the Oort Cloud was created after the birth of the Solar System. Oort imagined asteroids being flung out of the Solar System by close encounters with Jupiter. He was wrong. Asteroids are not like icy comets. There is another source: the Kuiper Belt.

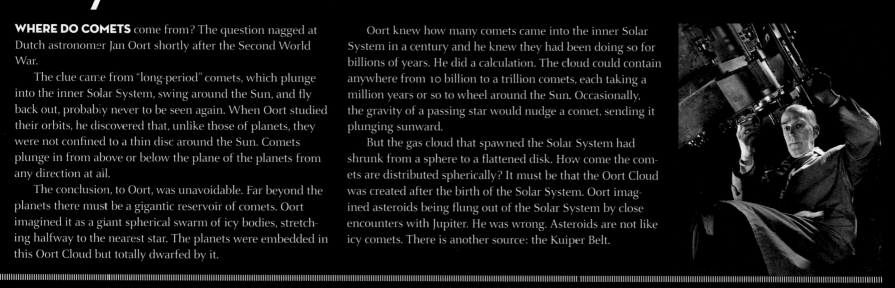

▼ Jan Oort guessed that our Solar System is embedded in a giant cloud of comets extending halfway to the next star.

◀ Comet McNaught became the brightest comet for 40 years when it swung by the Sun in 2007.

Comets

IMAGINE YOU ARE living before the age of science. The world appears chaotic, unreliable, scary. The heavens, by contrast, are regular, predictable, safe. Every night, the stars burn down, perfect and unchanging. The planets move against the stars, but their movements are at least predictable. The heavens will never surprise you. They will never shock. And then, one night, in this reliable perfection there appears a heart–stoppingly terrifying apparition–a glowing globe trailing hellfire across the sky.

Comets instilled terror. They were considered harbingers of catastrophe, bringers of doom, disease, and death. The truth is less dramatic, and far more interesting.

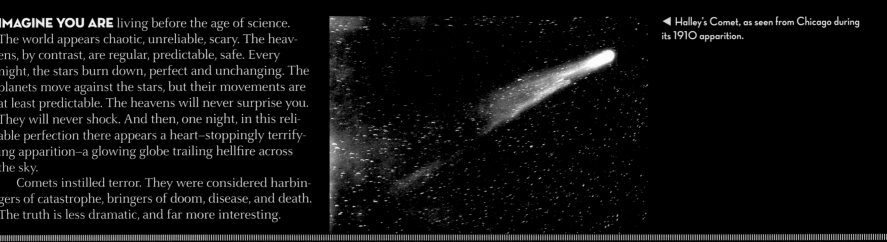

◀ Halley's Comet, as seen from Chicago during its 1910 apparition.

TOTAL NUMBER OF COMETS

200 (C/1973 N1 (Sandage), C/1991 L4 (Helin–Alu), C/1996 P2 (Russell–Watson), C/2006 YC (Catalina–Christensen), C/2002 U2 (LINEAR), C/2006 Q1 (McNaught), C/1970 N1 (Abe), C/1993 K1 (Shoemaker–Levy), C/1983 R1 (Shoemaker), C/1994 N1 (Nakamura–Nishimura–Machholz), C/1973 E1 (Kohoutek), C/2005 B1 (Christensen), C/1862 X1 (Bruhns), C/1980 L1 (Torres), C/2006 S3 (LONEOS), C/1975 V2 (Bradfield), C/1937 C1 (Whipple), C/1989 Q1 (Okazaki–Levy–Rudenko), C/1983 O2 (IRAS), C/1914 F1 (Kritzinger), C/2005 E2 (McNaught), C/1991 Y1 (Zanotta–Brewington), C/2007 W3 (LINEAR), C/1962 C1 (Seki–Lines), C/2007 W1 (Boattini), C/1997 N1 (Tabur), C/2006 P1 (McNaught), C/1987 W2 (Furuyama), C/1998 U1 (LINEAR), C/1999 K8 (LINEAR), C/1954 O2 (Baade), C/1999 S4 (LINEAR), C/2006 W3 (Christensen), C/1981 G1 (Elias), C/1890 F1 (Brooks), C/1913 Y1 (Delavan), C/1979 M1 (Bradfield), C/1915 C1 (Mellish), C/1999 J2 (Skiff), C/1973 W1 (Gibson), C/2006 E1 (McNaught), C/1997 J2 (Meunier–Dupouy), C/1998 W3 (LINEAR), C/1892 F1 (Denning), C/1989 Y1 (Skorichenko–George), C/1919 Q2 (Metcalf), C/2002 J5 (LINEAR), C/1973 A1 (Heck–Sause), C/2003 O1 (LINEAR), C/2005 L3 (McNaught), C/1960 M1 (Humason), C/2006 OF2 (Broughton), C/2003 S3 (LINEAR), C/1902 X1 (Giacobini), C/1906 B1 (Brooks), C/2007 F1 (LONEOS), C/1979 M3 (Torres), C/1975 N1 (Kobayashi–Berger–Milon), C/1995 Y1 (Hyakutake), C/1954 M2 (Kresak–Peltier), C/1978 R3 (Machholz), C/2006 M4 (SWAN), C/2007 D1 (LINEAR), C/1925 W1 (Van Biesbroeck), C/2006 VZ13 (LINEAR), C/2006 K1 (McNaught), C/1930 E1 (Beyer), C/1986 P1–A (Wilson), C/1986 P1 (Wilson), C/1900 O1 (Borrelly–Brooks), C/1895 W1 (Perrine), C/1984 K1 (Shoemaker), C/1988 L1 (Shoemaker–Holt–Rodriquez), C/1941 K1 (van Gent), C/1987 F1 (Torres), C/1949 K1 (Johnson), C/2003 G1 (LINEAR), C/2007 G1 (LINEAR), C/1999 Y1 (LINEAR), C/1991 F2 (Helin–Lawrence), C/1959 Y1 (Burnham), C/2003 K4 (LINEAR), C/1989 W1 (Aarseth–Brewington), C/2000 Y1 (Tubbiolo), C/1976 D2 (Schuster), C/2002 O7 (LINEAR), C/2006 K3 (McNaught), C/2004 D1 (NEAT), C/2001 B2 (NEAT), C/1993 Q1 (Mueller), C/1922 U1 (Baade), C/2001 G1 (LONEOS), C/1978 G2 (McNaught–Tritton), C/1847 J1 (Colla), C/2007 JA21 (LINEAR), C/1999 J4 (LINEAR), C/1921 E1 (Reid), C/1904 Y1 (Giacobini), C/1932 M1 (Newman), C/2000 K1 (LINEAR), C/1991 C3 (McNaught–Russell), C/1947 O1 (Wirtanen), C/1948 E1 (Pajdusakova–Mrkos), C/1946 U1 (Bester), C/2001 N2 (LINEAR), C/2000 A1 (Montani), C/1997 D1 (Mueller), C/1998 M3 (Larsen), C/1987 A1 (Levy), C/1954 O1 (Vozarova), C/1885 X1 (Fabry), C/2001 B1 (LINEAR), C/1952 W1 (Mrkos), C/1990 M1 (McNaught–Hughes), C/1880 G1 (Schaeberle), C/1892 Q1 (Brooks), C/1983 O1 (Cernis), C/1996 J1–B (Evans–Drinkwater), C/2000 WM1 (LINEAR), C/1972 U1 (Kojima), C/1950 K1 (Minkowski), C/1911 S3 (Beljawsky), C/1947 S1 (Bester), C/1907 E1 (Giacobini), C/1904 H1 (Brooks), C/1885 X2 (Barnard), C/1988 B1 (Shoemaker), C/2006 L2 (McNaught), C/1935 Q1 (Van Biesbroeck), C/1863 T1 (Baeker), C/1997 A1 (NEAT), C/1987 H1 (Shoemaker), C/1847 T1 (Mitchell), C/2003 WT42 (LINEAR), C/2007 U1 (LINEAR), C/1978 A1 (West), C/1968 N1 (Honda), C/1999 K5 (LINEAR), C/2000 H1 (LINEAR), C/2002 E2 (Snyder–Murakami), C/1925 G1 (Orkisz), C/1898 L1 (Coddington–Pauly), C/2003 T4 (LINEAR), C/1925 F1 (Shajn–Comas Sola), C/1999 T3 (LINEAR), C/2003 A2 (Gleason), C/1996 N1 (Brewington), C/1977 D1 (Lovas), C/1886 T1 (Barnard–Hartwig), C/1956 F1–A (Wirtanen), C/1942 C1 (Whipple–Bernasconi–Kulin), C/1932 M2 (Geddes), C/2006 S2 (LINEAR), C/1955 G1 (Abell), C/1971 E1 (Toba), C/1912 R1 (Gale), C/1989 X1 (Austin), C/1974 V1 (van den Bergh), C/1925 V1 (Wilk–Peltier), C/1999 T2 (LINEAR), C/1986 P1–B (Wilson), C/1976 U1 (Lovas), C/1996 E1 (NEAT), C/2002 R3 (LONEOS), C/2005 Q1 (LINEAR), C/1990 K1 (Levy), C/1888 R1 (Barnard), C/1946 C1 (Timmers), C/1946 P1 (Jones), C/1948 T1 (Wirtanen), C/1953 T1 (Abell), C/2005 A1–A (LINEAR), C/2001 Q4 (NEAT), C/1947 Y1 (Mrkos), C/1987 Q1 (Rudenko), C/1978 H1 (Meier), C/1999 N4 (LINEAR), C/2007 K4 (Gibbs), C/1908 R1 (Morehouse), C/1942 C2 (Oterma), C/2001 RX14 (LINEAR), C/1956 R1 (Arend–Roland), C/2002 T7 (LINEAR), C/1996 J1–A (Evans–Drinkwater), C/1900 B1 (Giacobini), C/2004 H6 (SWAN), C/1959 Q1 (Alcock), C/1849 G2 (Goujon), C/1853 L1 (Klinkerfues), C/1980 R1 (Russell), C/1999 S2 (McNaught–Watson), C/2004 B1 (LINEAR), C/2005 EL173 (LONEOS), C/1999 U1 (Ferris), C/1896 G1 (Swift), C/1999 H3 (LINEAR), C/2000 OF8 (Spacewatch), C/1914 M1 (Neujmin), C/2005 K1 (Skiff), and C/2002 B3 (LINEAR))

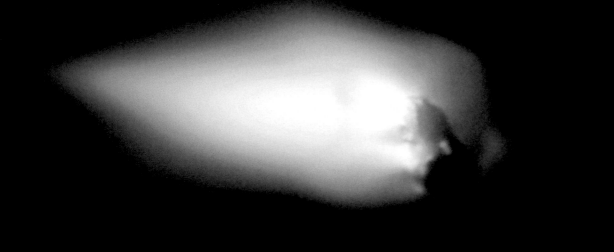

◀ The European Space Agency's Giotto spacecraft flew within 600 km of the nucleus of Halley's Comet on March 13, 1986. The image above shows jets emerging from at least two areas of Halley's nucleus.

A trillion comets

ON MARCH 14, 1986, the European Space Agency's Giotto space probe plunged into the head of Halley's Comet. No one knew whether it would survive the vicious sand-blasting from dust particles. But it did. And there, looming out of the mist, frozen in time since the dawn of the Solar System, was something nobody had seen before: a cometary nucleus.

A mere 15 kilometers long and shaped like a peanut, the surprise was how dark it was–blacker even than coal. But, here and there where the heat of the Sun had burned through like a blow-torch the filthy crust was punctuated by patches of pristine white snow. From these patches great jets of vapor stabbed outward, the ultimate source of the comet's multimillion-mile-long tail.

No comet has captured the public's imagination like Halley's. Its appear-ance in 1066 is immortalized in the Bayeux Tapestry commemorating the Norman invasion of England. The 14th-century Italian artist Giotto di Bondone even depicted the comet as the star of Bethlehem, hence the name of ESA's comet probe. In 1705, Newton's friend Edmund Halley deduced that comets seen in 1456, 1531, 1607, and 1682 were in fact one and the same comet moving around the Sun in a highly elongated orbit every 75 to 76 years. Halley, who died at 85 in 1742, predicted its return in 1758. That prediction's success led to the name Halley's Comet.

Halley's visits have been observed since at least 240 B.C. Currently it is in-visible, heading out to the farthest point in its orbit, beyond Neptune. From there it will begin its plunge back toward the warmth of the Sun, returning to Earth's skies in 2061.

▲ Halley's Comet is shown in the Bayeux Tapestry, which commemorates the Norman invasion of Britian of 1066.

Comet tails

IMAGINE A MOSQUITO leaving a vapor trail more than a kilometer long. This is the way it is for comets. Although a comet's tail is sometimes long enough to span the gulf between the Sun and Earth, the nucleus responsible for the spectacular show is often only a few kilometers across.

Cometary nuclei are "dirty snowballs," loose aggregates of ice and dust left over from the birth of the Solar System. Usually, they orbit far from the Sun—either in the distant Oort Cloud or the nearer Kuiper Belt—too small and cold to be visible. But when a nudge from a passing star or nearby body sends them plunging sunward, all hell breaks loose.

As they approach the Sun, swing around, and head back out, solar heat causes the ices—compounds such as methane, ammonia, water, and carbon dioxide—to boil off, creating a giant cloud, or "coma," around the nucleus. The million-mile-an-hour hurricane of the solar wind blows the gases into a tail, which shines from reflected sunlight, while the "pressure" of light from the Sun blows the dust particles into a second tail.

The two tails—essentially solar windsocks—always point away from the Sun. So bizarrely, when the comet heads back to the edge of the Solar System, its tails go ahead.

If a comet becomes trapped in the inner Solar System, its repeated passes by the Sun each boil off about a meter of its surface and eventually wear it away until only a rocky core is left. That is, if the Sun's gravity does not break it up entirely.

◀ Comet Hale–Bopp is a long–period comet that made a spectacular visit to the inner solar system in 1997, displaying a well separated blue plasma tail.

▲ Halley develops a long tail as it passes through the inner solar system every 75–76 years, making it visible to the naked eye.

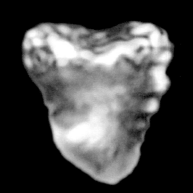

▲ The Stardust mission successfully returned samples of material from Comet Wild 2 to Earth. This beautiful heart–shaped dust particle is about 0.01 mm across.

▲ Aerogel was used on the Stardust mission to catch and hold delicate samples of comet dust. This magnified image shows where a comet particle has penetrated the aerogel and exploded.

◀ The nucleus of Comet Schwasmann-Wach-mann 3 was torn apart as it swung round the Sun in 2006.

The meteor connection

ALL NIGHT IT rained fire. At the peak of the meteor storm of November 12, 1833 more than 1,000 shooting stars a minute were streaking down the sky from the constellation of Leo. Many in North America thought it was Judgment Day.

It was not until the late 19th century that a connection was made with comets. Giovanni Schiaparelli, who would find channels, or "canali," on Mars, noted that the orbits of the tiny dust particles responsible for a meteor display were often the same as that of a comet.

Comets are dirty snowballs. As they approach the Sun, their ices boil off into space, carrying with them tiny dust particles. These form a stream along the comet's orbit. If the Earth passes through the stream, these tiny dust particles or micrometeorites stab down through the atmosphere,

friction with the air causing them to burn up as meteors, or shooting stars.

Many comets trapped in the inner Solar System have dust streams spread out along their orbits. Once a year, as the Earth passes through a particular stream, we get a meteor shower. The Leonids, associated with Comet Tempel-Tuttle, occur in mid-November. The 1833 spectacular was just a particularly intense Leonid shower.

Occasionally, the Earth passes through a cometary dust stream just after the comet has made its closest approach to the Sun and a large amount of fresh dust has come off the cometary nucleus. Then we get not just a shower but a meteor storm. On a typical night there are about 10 meteors an hour. In a shower it can be 100. But in a storm the number can be huge.

Death from the sky

ON EARTH, DEATH comes with metronomic regularity. According to the fossil record, there is a great die-off of species every 27 million years.

The obvious culprit is a comet impact. A small comet fragment about the size of a terrace of houses disintegrated above Siberia's Tunguska river in 1908. The blast wave flattened more than 2,000 square kilometers of forest. But why should comet impacts be regular?

At regular intervals something stirs up the Oort Cloud, sending a flurry of comets sunward, some on a collision course with Earth. In 1984 David Raup and John Sepkoski made the extraordinary suggestion that the Sun, like most stars, is actually a binary. What if it has a super-faint companion in a highly elongated orbit that, once every 27 million years, brings it near the Sun and shakes up the Oort Cloud?

This Nemesis star has never been found. And in 2010 researchers discovered that the regularity of the fossil record is actually too regular. The gravitational tug of nearby stars would cause the orbital period of Nemesis to vary, which is not what the fossil record shows.

Another explanation, proposed by astronomers Bill Napier and Victor Clube, is that the 27-million-year period is related to the Solar System's motion around the Galaxy. As it orbits the center, it also oscillates up and down, taking it periodically through the plane of the Galaxy, where massive star-forming regions known as Giant Molecular Clouds are concentrated. Their gravity, according to Napier and Clube, may stir up the Oort Cloud every 27 million years.

◀ The explosion of a body, possibly a comet, high above Siberia in 1908, flattened an estimated 80 million trees.

◀ The horsehead nebula is part of a cold, dark cloud of gas and dust where stars—and comets—are born.

▲ The nucleus of Comet Wild 2 is about 5 kilometers in diameter. It was visited by the Stardust probe in January 2004.

▶ The nucleus of Comet Holmes is but a bright speck at the center of its immense coma in this infrared image from the Hubble Space Telescope.

Were our comets stolen?

THE RECEIVED WISDOM is that comets are the icy debris left over from the birth of the Solar System. But the received wisdom may be wrong. Remarkably, many famous comets such as Halley's Comet and Comet Hale-Bopp may come from other solar systems.

This extraordinary idea might explain a puzzle about comets: there are too many of them. Comets are thought to be chunks of icy rubble catapulted out of the inner Solar System by close encounters with the gravity of embryonic giant planets like Jupiter. But this process would have filled the Oort Cloud with only about 10 percent of the comets we infer are out there.

For the other 90 percent, we may have to look to other stars, according to a team led by Hal Levison of the Southwest Research Institute in Boulder, Colorado. In 2010 they pointed out that

the Sun formed in a "stellar nursery," where it rubbed shoulders with hundreds of other stars. Most of the chunks of icy debris around each of the stars were flung outward in encounters with giant planets, becoming free-floating members of the star cluster.

Crucially, when massive, luminous stars switched on, they blew away the placental gas and dust and broke apart the young star cluster. As the Sun sailed away, its gravity would have vacuumed up comets as it went. This process turns out to be surprisingly efficient, says Levison's team, and the Sun could easily have acquired 90 percent of its comets this way.

If the scenario is right, we do not have to go to other stars to sample material from other stars. It is already in our backyard.

The life plague

LIFE GOT STARTED on Earth pretty much as soon as the planet had cooled down enough. So surely biology gets going easily? Yet life has been impossible to create from nonlife in the laboratory. These two facts can be reconciled, according to Chandra Wickramasinghe and the late Sir Fred Hoyle, if life were seeded from space. In their scheme, a modern twist on an old idea called "panspermia," the bodies that transfer life between stars, and planets, are comets.

It works like this. Interstellar gas clouds are graveyards for desiccated bacteria. When such a cloud shrinks, the bacteria are destroyed by heat if they end up in newly forming stars and planets. But not in icy comets.

Initially a comet is melted by the decay of radioactive elements such as aluminum-26. In the comet's liquid center, any bacteria that are viable multiply explosively. Later, when a comet falls sunward, material containing bacteria boils off its surface into space. It is material like this, they say, that was swept up by the Earth around 4 billion years ago and seeded the planet with life.

According to Wickramasinghe and Hoyle, terrestrial bacteria, wafted on air currents to the top of the atmosphere, escape into space. They are then driven by the pressure of sunlight back into interstellar space. Thus the great cosmic cycle of life is completed.

But how did it begin? Wickramasinghe and Hoyle do not know. The point is, in their scheme, it does not matter if the emergence of life is unbelievably unlikely. It needs only to have started once, somewhere in the Galaxy. Thereafter, by means of comets, it could have spread anywhere else—including to Earth.

▼ The 2-kilometer-long nucleus of comet Hartley 2 was observed from a distance of about 700 kilometers during the brief flyby of Deep Impact probe in November 2010.

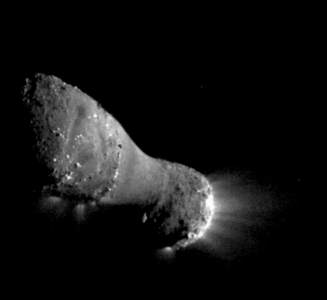

▼ The moment of impact, as a small probe is vaporized by its collision with Comet Temple 1 on July 4, 2005

▶ The Shell (or coma) of Comet Holmes briefly became the largest object in the Solar System as it rapidly expanded in October 2007. At that time it was on the far side of the Sun, so its tail appeared foreshortened when seen from Earth.

Credits

Pg. 8: Planetary Visions Pg. 10: Joe Zeff Design Pg. 12: NASA/Planetary Visions Ltd. Pg. 14: Planetary Visions Pg. 15: Mosaic of two-minute exposures using a Nikon D300 digital camera with a 14mm lens on an equatorial mount, taken at Cerro Paranal, Chile. (Bruno Gilli / ESO) Pg. 16: (top) ©York Films; (bottom) Combination of hydrogen light and oxygen light observations by the Advanced Camera for Surveys on the Hubble Space Telescope. (NASA / ESA / Hubble Heritage Project (STScI/AURA) / M Livio / N Smith, University of California, Berkeley) Pg. 17: (bottom) Infrared image from the 4.1-meter Visible and Infrared Survey Telescope for Astronomy (VISTA) at the European Southern Observatory, Paranal, Chile. (ESO / J Emerson / VISTA / Cambridge Astronomical Survey Unit) Pg. 18: Near infrared image of the star 1RXS J160929.1-210524 in J-, H- and K-bands taken using adaptive optics on the 8.1-meter Gemini North telescope in Hawaii, in 2008. (Gemini Observatory) Pg. 19: (top) Planetary Visions; (bottom) Image taken with the Wide Field and Planetary Camera 2 on the Hubble Space Telescope in December 1993. (C R O'Dell, Rice University / NASA) Pg. 20: 45-minute exposure from a digital camera using a 10mm lens and equatorial mount, taken at Paranal, Chile, home of ESO's Very Large Telescope. (ESO/Y Beletsky) Pg. 21: Photograph of American astronaut Bruce McCandless ©NASA Pg. 22: (top) Natural color image from the Multi-angle Imaging Spectro-Radiometer (MISR) on NASA's Terra spacecraft, showing an area about 380 km across. (NASA/GSFC/LaRC/JPL, MISR Team); (bottom) Photograph of Italian physicist Enrico Fermi ©U.S. Department of Energy Pg. 23: Planetary Visions Pg. 24: Planetary Visions
Pg. 28: (bottom, left) Image taken by the Solar Optical Telescope on Japan's Hinode satellite on November 20, 2006. (Hinode JAXA/NASA); (bottom, right) This image was taken at a wavelength of 171 Angstroms by the Extreme ultraviolet Imaging Telescope (EIT) on the Solar and Heliospheric Observatory satellite (SOHO). (ESA / NASA)
Pg. 29: (top, left) This image from the SOHO spacecraft combines data from two instruments: the LASCO coronograph, which blocks light from the Sun's bright disc to observe its faint corona, and the EIT, which observes the Sun's surface in ultraviolet light. (ESA / NASA); (top,center) Images from the Extreme Ultra Violet Imager (EUVI) on NASA's two Solar TErrestrial RElations Observatory (STEREO) satellites. (NASA/JPL-Caltech/NRL/GSFC); (top, right) Corona photo taken during the eclipse in Australia on December 3, 2002. Sun's photosphere image from the Extreme ultraviolet Imaging Telescope (EIT) on the Solar and Heliospheric Observatory (SOHO). (NASA/ESA); (bottom, left) Multi-wavelength ultraviolet image from the Extreme ultraviolet Imaging Telescope (EIT) on the Solar and Heliospheric Observatory satellite (SOHO). (ESA / NASA); (bottom, right) Multi-wavelength ultraviolet image from the Atmospheric Imaging Assembly (AIA) on the Solar Dynamics Observatory satellite. (NASA/SDO/AIA) Pg. 30: (top and bottom) Ground-based photograph using a 92mm refracting telescope with a hydrogen alpha filter (656nm) and industrial CCD camera. Multiple video frames were stacked and averaged to sharpen the image. (©Alan Friedman / avertedimagination.com) Pg. 31: (top) Touch Press; (bottom) This image was taken in extreme ultraviolet light by the STEREO space telescope on August 25, 2010. (NASA) Pg. 32: Extreme ultraviolet image taken on April 21, 2010 by the Atmospheric Imaging Assembly (AIA) on NASA's Solar Dynamics Observatory spacecraft. (NASA/SDO) Pg. 33: Image of sunspot AR NOAA 1084 taken by the 1.6 meter New Space Telescope at Big Bear Solar Observatory on July 2, 2010 using a TiO filter at 706nm. To get such a detailed view, the telescope's mirror is distorted in real time to compensate for the effects of atmospheric disturbances - a technique known as adaptive optics. (©BBSO/NJIT) Pg. 34: Photograph of Carrington event magnometer readings ©British Geological Survey (NERC) Pg. 36: (top) SOHO rotation courtesy ESA/NASA; (bottom) Image taken using the PSPT/CaIIK camera at the Mona Loa Solar Observatory March 28, 2001. (NASA/Goddard Space Flight Center Scientific Visualization Studio) Pg. 37: (top) STEREO coronograph movie courtesy NASA; (bottom) Touch Press Pg. 40: (top) MDIS Wide Angle Camera image from the MESSENGER spacecraft, sensitive to 11 wavelength bands between 400 and 1050nm. The natural color image on the left uses color filters at 480nm, 560nm and 630nm. (NASA/Johns Hopkins University Applied Physics Laboratory/Carnegie Institution of Washington); (bottom) Image from the Narrow Angle Camera (NAC) of the Mercury Dual Imaging System (MDIS) on the MESSENGER spacecraft. (NASA/Johns Hopkins University Applied Physics Laboratory/Carnegie Institution of Washington) Pg. 41: (top) False color image from the Wide Angle Camera (WAC) of the Mercury Dual Imaging System (MDIS) on the MESSENGER spacecraft. Infrared (1000nm), far red (700nm) and violet (430nm) filters were used for this view. (NASA/Johns Hopkins University Applied Physics Laboratory/Carnegie Institution of Washington); (center) Combination of images from the Wide Angle Camera (WAC) and the high resolution Narrow Angle Camera (NAC) of the Mercury Dual Imaging System (MDIS) on the MESSENGER spacecraft. (NASA/Johns Hopkins University Applied Physics Laboratory/Carnegie Institution of Washington) Pg.42: Image from the Solar Optical Telescope on the Japanese solar observing satellite Hinode. (Hinode JAXA/NASA/PPARC) Pg. 43: Planetary Visions Pg. 44: (top) Artist's impression of the MErcury Surface, Space ENvironment, GEochemistry, and Ranging spacecraft (MESSENGER) in orbit around Mercury. (NASA/Johns Hopkins University Applied Physics Laboratory/Carnegie Institution of Washington); (bottom) Planetary Visions Pg. 45: A three-image mosaic from the Wide Angle Camera (WAC) of the Mercury Dual Imaging System (MDIS) on the MESSENGER spacecraft. (NASA/Johns Hopkins University Applied Physics Laboratory/Carnegie Institution of Washington) Pg. 48: (bottom, left) Ultraviolet image from Pioneer Venus Orbiter, taken on February 26, 1979. (NASA/JPL); (bottom, right) Image from Venus Express combining VIRTIS infrared imagery at a wavelength of 1.7 microns for the night side, with visible/ultraviolet imagery for the day side. (ESA/CNR-IASF, Rome, Italy, and Observatoire de Paris, France) Pg. 49: (top, left) Infrared images at a wavelength of 1.7 microns, taken by Venus Express on July 22, 2006. (ESA/VIRTIS/INAF-IASF/Obs. de Paris-LESIA); (top, right) Ultraviolet VIRTIS image at a wavelength of 380 nm, taken by Venus Express from a distance of 190,000 km. (ESA/VIRTIS/INAF-IASF/Obs. de Paris-LESIA); (center, left) False color infrared image at 1.27 microns (blue) and 1.7 microns (yellow), taken by the VIRTIS instrument on Venus Express. (ESA/VIRTIS/INAF-IASF/Obs. de Paris-LESIA); (center, right) Simulated color image based on Magellan Imaging Radar data and colors seen by the Venera 13 lander. (NASA/JPL); (bottom, left) Panoramic Telephotometer image from Venera 13 lander at latitude 7.5° South, longitude 303.0° East, using dark blue, green and red filters. (NASA/GSFC); (bottom, right) Image from Pioneer Venus Orbiter. (NASA/ARC) Pg. 50: (top) Planetary Visions; (bottom) Simulated view from 4 km above the surface, with a height exaggeration of 6 times, based on Magellan Imaging Radar mosaic and Radar Altimeter data. (NASA/JPL) Pg. 51: (top) Magellan Imaging Radar image mosaic. The surface appears bright or dark according to its roughness, with rougher surfaces reflecting more radar energy back to the satellite, so appearing brighter. (NASA/JPL); (bottom) Magellan Imaging Radar image. (NASA/JPL)
Pg. 52: (top) Image from the Transition Region And Coronal Explorer (TRACE) spacecraft, taken at 05:34 Universal Time on June 8, 2004. (NASA/Lockheed Martin Solar Astrophysics Laboratory); (bottom) H-alpha filter image from the Swedish 1 meter Solar Telescope. (Institute for Solar Physics of the Royal Swedish Academy of Sciences)
Pg. 53: Planetary Visions Pg. 56: (center) Large format photograph taken with a 70 mm camera from the Apollo 17 Command Module on December 7, 1972. (NASA-JSC); (bottom, left) Composite image based on ultraviolet data from the Imager for Magnetopause-to-Aurora Global Exploration (IMAGE) spacecraft, taken on 11th September 2005. (NASA); (bottom, right) Photo taken using a Nikon D3 digital camera from the International Space Station during Expedition 23, on May 29, 2010. (NASA-JSC) Pg. 57: (top, left) Photo from a Nikon D2Xs digital camera using a 200 mm lens, taken from the International Space Station during Space Shuttle mission 125, on May 13, 2009. (NASA-JSC); (top, right) Photo from a Kodak DCS760C digital camera using a 30 mm lens, taken from the International Space Station during Expedition 5, on July 20, 2006. (NASA-JSC); (bottom, left) False color infrared image from the Advanced Spaceborne Thermal Emission and Reflection Radiometer (ASTER) on NASA's Terra satellite. (NASA/GSFC/METI/ERSDAC/JAROS, and U.S./Japan ASTER Science Team); (bottom, right) Natural color image from the Enhanced Thematic Mapper (ETM+) on Landsat 7. (NASA / Serge Andrefouet, University of South Florida) Pg. 58: (top) Digital air photo orthoimage mosaic. (New York State); (bottom) Natural color image from the Enhanced Thematic Mapper (ETM+) on Landsat 7, taken on February 21, 2000. (Landsat Science Team / NASA-GSFC) Pg. 59: (top) Natural color image from the Advanced Land Imager on the Earth Observing 1 (EO1) satellite, taken on September 6, 2010. (NASA); (bottom) Natural color image from the Advanced Land Imager on the Earth Observing 1 (EO1) satellite, taken on September 4, 2010. (NASA); (bottom, right) Natural color image from the Advanced Spaceborne Thermal Emission and Reflection Radiometer (ASTER) on NASA's Terra satellite. (NASA/GSFC/METI/ERSDAC/JAROS, and U.S./Japan ASTER Science Team)
Pg. 60: (top, left) Photo from a Nikon D2Xs digital camera using a 400 mm lens, taken from the International Space Station during Expedition 20, on June 12, 2009. (NASA-JSC); (bottom) False color infrared image from the Advanced Spaceborne Thermal Emission and Reflection Radiometer (ASTER) on NASA's Terra satellite. (NASA/GSFC/METI/ERSDAC/JAROS, and U.S./Japan ASTER Science Team); (bottom, left) Photo from a Nikon D2Xs digital camera using an 80 mm lens, taken from the International Space Station during Expedition 24, on August 22, 2010. (NASA-JSC) Pg. 61: (bottom) Natural color image from the Enhanced Thematic Mapper (ETM+) on Landsat 7, taken on March 19, 2002. (NASA / UMD Global Land Cover Facility)
Pg. 62: (top) Photo from a Kodak DCS760C digital camera using a 400 mm lens, taken from the International Space Station during Expedition 13, on July 20, 2006. (NASA-JSC); (center) Images show the same part of North America at 4:15pm on March 25, 1999, at visible, mid-infrared and far-infrared wavelengths (0.65, 6.7 and 11 microns). Images from the GOES Imager on the Geostationary Operational Environmental Satellite (GOES-8). (NOAA-NASA GOES Project); (bottom, left) Image from the Moderate Resolution Imaging Spectroradiometer (MODIS) on the TERRA research satellite (NASA - MODIS Science Team); (bottom, right) Photo from a Nikon D2Xs digital camera using a 28-70 mm zoom lens set at 48 mm, taken from the International Space Station during Expedition 20, on October 6, 2009. (NASA-JSC) Pg. 63: (top) Photo from a Nikon D2Xs digital camera using an 800 mm lens, taken from the International Space Station during Expedition 18, on January 6, 2009. (NASA-JSC); (bottom) Natural color image from the Advanced Spaceborne Thermal Emission and Reflection Radiometer (ASTER) on NASA's Terra satellite. (NASA/GSFC/METI/ERSDAC/JAROS, and U.S./Japan ASTER Science Team) Pg. 64: (top) 3D visualization using data from the Microwave Limb Sounder and the Total Ozone Mapping Spectrometer, on the Aura and Earth Probe satellites. (NASA - GSFC Scientific Visualization Studio); (bottom) Greek mathematician Eratosthenes courtesy of Wikimedia Commons Pg. 65: (top, left) Natural color image from the Advanced Spaceborne Thermal Emission and Reflection Radiometer (ASTER) on NASA's Terra satellite, taken in September 2006. (NASA/GSFC/METI/ERSDAC/JAROS, and U.S./Japan ASTER Science Team); (top, right) Natural color image from the Enhanced Thematic Mapper (ETM+) on Landsat 7, taken in June 2001. (NASA / UMD Global Land Cover Facility); (center, left) Photo from a Kodak DCS760C digital camera taken from the International Space Station on October 14, 2002. (NASA-JSC); (center, right) Natural color image from the Enhanced Thematic Mapper (ETM+) on Landsat 7, taken in August 2001. (NASA / UMD Global Land Cover Facility); (bottom, left) Photo from a Nikon D3S digital camera using a 16 mm lens, taken from the International Space Station during Expedition 25, on October 28, 2010. (NASA-JSC); (bottom, right) Photo from the departing Space Shuttle Atlantis, taken on May 23, 2010. (NASA-JSC) Pg. 68: (bottom, left) Enhanced color image taken with violet and near-infrared filters by the Galileo probe. (NASA/JPL/USGS); (bottom, right) Photo from the departing Space Shuttle Atlantis, taken on May 23, 2010. (NASA-JSC) Pg. 69: (top, left) Astronaut photo from the International Space Station, taken with a Nikon D2Xs electronic stills camera, 200mm lens, 1/1000 sec exposure, image number ISS024-E-013819. (NASA/JSC); (top, right) Combination of several 1/10 second exposures through a near-infrared filter at 856 nm. (ESO); (center, left) Apollo 11 70mm Hasselblad color frame number AS11-40-5877. (NASA/LPI); (center, right) A mosaic of

500 images from the Clementine orbiter, taken using 415 nm, 750nm and 1000 nm filters. (NASA/JPL/USGS); (bottom, left) Lunar Reconnaissance Orbiter (LRO) Narrow Angle Camera (NAC) image. (NASA/GSFC/Arizona State University); (bottom, right) Image from the Japanese Space Research Agency's Kaguya satellite, using the HDTV-WIDE camera. (Courtesy of JAXA/NHK) Pg. 70: (top) Apollo 11 70mm Hasselblad color frame number AS11-40-5903. (NASA); (center, left) Apollo 11 70mm Hasselblad color frame number AS11-40-5877. (NASA/LPI); (center, right) Photograph of geologist and astronomer Gene Shoemaker ©Roger Ressmeyer/CORBIS; (bottom) Apollo 17 panorama courtesy NASA-JSC Pg. 71: (top) Apollo 11 70mm Hasselblad color frame number AS11-40-5927. (NASA/LPI) Pg. 72: (top) Photograph of American astronaut Eugene Cernan courtesy of PVL/NASA; (center) Apollo Lunar Surface Closeup Camera (ALSCC) 35mm stereoscopic image pair from Apollo 11. (NASA/LPI); (bottom, left) Polarized light microscope image from Apollo 11 rock sample 10003. (NASA/LPI); (bottom, right) Photographs of rocks returned form the Apollo landings. (Lunar and Planetary Institute) Pg. 73: (top) Geological map from USGS Miscellaneous Investigation Series Map I-948. Moon formation ©York Films; (bottom, left and right) Image from the Lunar Reconnaissance Orbiter (LRO) Narrow Angle Camera (NAC). (NASA/GSFC/Arizona State University) Pg. 74: Planetary Visions Pg. 75: (top, left) Photograph of the Laser Ranging Facility at the Geophysical and Astronomical Observatory at NASA's Goddard Space Flight Center, Greenbelt, Maryland. (NASA/GSFC); (top, right) Apollo 14 70mm Hasselblad color frame number AS14-67-9385. (NASA/JSC); (bottom) Touch Press Pg. 76: (top) Duhoux ESO Pg. 77: (top) Apollo 8 70mm Hasselblad color frame number AS8-14-2383. (NASA/LPI) (bottom) @York Films Pg. 80: (bottom, far left and left) Natural color image from the Wide Field /Planetary Camera on the Hubble Space Telescope. (NASA / ESA / The Hubble Heritage Team, STScI/AURA); (bottom, center) Natural color image from the Wide Field /Planetary Camera on the Hubble Space Telescope. (NASA/James Bell, Cornell Univ/ Michael Wolff, Space Science Inst /The Hubble Heritage Team, STScI/AURA); (bottom, right) Visible/infrared color image from the High Resolution Imaging Science Experiment on the Mars Reconnaissance Orbiter. (NASA/JPL/University of Arizona) Pg. 81: (top, left) Visible/infrared color image from the High Resolution Imaging Science Experiment on the Mars Reconnaissance Orbiter. (NASA/JPL/University of Arizona); (top, right) False color image from the Panoramic Camer on the Mars Reconnaissance Rover Opportunity. (NASA/JPL/Cornell); (center, left) Natural color image from the High Resolution Imaging Science Experiment on the Mars Reconnaissance Orbiter. (NASA/JPL/University of Arizona); (center, right) False-color image from the Thermal Emission Imaging System on the Mars Odyssey satellite. (NASA/JPL/ASU); (bottom) Enhanced color image from the High Resolution Imaging Science Experiment on the Mars Reconnaissance Orbiter. (NASA /JPL-Caltech /University of Arizona) Pg. 82: Photographs of Percival Lowell's sketches of Mars courtesy of Wikimedia Commons Pg. 83: (main image) Natural color image from the vidicon camera on the Viking 1 Orbiter. (NASA/JPL); (top, left) Painting based on Viking Orbiter monochrome image mosaic. (NASA/Gordon Legg); (top,

right) Natural color mosaic based on Viking Orbiter imagery. (NASA/JPL/Planetary Visions); (bottom, left) Natural color mosaic based on Viking Orbiter imagery. (NASA/JPL/ USGS); (bottom, right) Planetary Visions Pg. 84: (top) Planetary Visions; (bottom, left) Natural color image mosaic from the Visual Imaging Subsystem on the Viking 2 Orbiter. (NASA/JPL/USGS); (bottom, right) Visible/infrared color image from the High Resolution Imaging Science Experiment on the Mars Reconnaissance Orbiter. (NASA/JPL-Caltech / University of Arizona) Pg. 85: (top, left and bottom, left) Simulated perspective view using imagery and height data from the High Resolution Stereo Camera on the European Space Agency's Mars Express. (ESA/DLR/FU Berlin, G. Neukum); (top, right) Image from the High Resolution Imaging Science Experiment on the Mars Reconnaissance Orbiter. (NASA/ JPL/University of Arizona); (bottom, right) Simulated perspective view using false color infrared imagery from the Compact Reconnaissance Imaging Spectrometer for Mars (CRISM), and height data from a stereo pair of images from the Context Camera, on the Mars Reconnaissance Orbiter. (NASA / JPL-Caltech/MSSS /JHU-APL/Brown Univ) Pg. 86: (top) Enhanced color image from the Panoramic Camera on the Mars Exploration Rover Opportunity. (NASA/JPL/ Cornell); (bottom) Mars Pathfinder panorama courtesy NASA/JPL Pg. 87: (top, left) Image from the Hazard Avoidance Camera on the Mars Exploration Rover Spirit (NASA/JPL/Cornell); (top, right) Image from the Microscopic Imager on the Mars Exploration Rover Spirit (NASA/JPL/Cornell); (center) Enhanced color image from the Panoramic Camera on the Mars Exploration Rover Opportunity. (NASA/JPL/Cornell) Pg. 88: (top) Comparison of images from the vidicon camera of the Viking 1 Orbiter (left) with the Mars Orbiter Camera on the Mars Global Surveyor (right). (NASA/JPL/Malin Space Science Systems); (bottom, left) Microscopic photo of extremophile bacteria courtesy of Professor Michael J. Daly, Uniformed Services University, Bethesda, Maryland; (bottom, right) Enhanced color image mosaic from the Panoramic Camera on the Mars Exploration Rover Spirit (NASA/JPL/Cornell) Pg. 89: Olympus Mons data courtesy NASA/JPL/USGS Pg. 91: (bottom) Enhanced color image from the High Resolution Image Science Experiment on the Mars Global Surveyor satellite, using blue/ green, red and near-infrared wavelengths. (NASA/JPL/ University of Arizona) Pg. 93: (top) Planetary Visions Pg. 96: Planetary Visions Pg. 98: Planetary Visions Pg. 99: Planetary Visions Pg. 100: (left) Touch Press; (right) Montage of asteroids visited by space probes Galileo, Rosetta and NEAR Shoemaker. (ESA/NASA/JHUAPL) Pg. 101: (left) False color image from NEAR-Shoemaker's Multi-Spectral Imager. (NASA/JPL/JHUAPL); (right) Montage of two images from the Multi-Spectral Imager on the NEAR Shoemaker probe. (NASA/JPL/JHUAPL); (bottom) Photograph of English musician and astrophysicist Brian May ©Imperial College London/Neville Miles Pg. 103: (top) NASA/JPL-Caltech/ UCLA/MPS/DLR/IDA; (bottom) Simulated natural color image taken in visible and ultraviolet light by the Hubble Space Telescope's Advanced Camera for Surveys. (NASA/ESA) Pg. 104: False color image taken in green and infrared light by NEAR-Shoemakers Multi-Spectral Imager. (NASA/ JPL/JHUAPL) Pg. 105: (top) Simulated view based on six images and a detailed surface model from NEAR-Shoemaker's

Laser Rangefinder. (NEAR Project/NLR/JHUAPL/ Goddard SVS/NASA); (bottom) False color image taken in green and infrared light by NEAR-Shoemaker's Multi-Spectral Imager. (NASA/JPL/JHUAPL) Pg. 107: Photograph of particles passing copyright Dr. Ruth Bamford Pg. 109: Simulated natural color view based on violet and infrared wavelength images (410nm, 756nm and 968nm) from the Solid State Imaging sensor on the Galileo space probe. (NASA/ JPL/USGS) Pg. 111: (top) Imagery from the Japanese Space Agency's probe Hyabusa. (Courtesy of JAXA); (center, right) Artist's impression of the Japanese Space Agency's probe Hyabusa touching down on its target. (Courtesy of JAXA); (bottom) Photograph of Japanese scientists (Courtesy of JAXA) Pg. 112: Planetary Visions Pg. 116: (top) Radio map at a frequency of 13.8 GHz (wavelength 2.2 cm) from the Cassini Orbiter's Imaging Radar in listen-only mode. (NASA/JPL); (center, left) Composite image in ultraviolet light from the Space Telescope Imaging Spectrograph, and visible light from the Wide Field/Planetary Camera 2, both instruments on the Hubble Space Telescope. (John Clarke, University of Michigan / NASA / ESA / Planetary Visions); (center, right) Composite image from the Chandra X-ray Observatory and Hubble Space Telescope. (X-ray: NASA/CXC/SwRI/R.Gladstone et al.; Optical: NASA/ESA/Hubble Heritage (AURA/STScI)); (bottom) False color image in ultraviolet light from the Space Telescope Imaging Spectrograph on the Hubble Space Telescope. (NASA / ESA / John T. Clarke, Univ. of Michigan) Pg. 117: (top, left) Enhanced color image from the Narrow Angle camera on Voyager 1. (NASA/JPL); (top, right) Natural color images from the Wide Field/Planetary Camera 2 on the Hubble Space Telescope. (NASA/ESA/A. Simon-Miller, Goddard Space Flight Center / N. Chanover, New Mexico State University / G. Orton, Jet Propulsion Laboratory); (bottom) Enhanced color image from the Narrow Angle camera on Voyager 1. (NASA/JPL) Pg. 118: (top) True color (left) and false color (right) image mosaics from the Solid State Imaging system on the Galileo Orbiter. (NASA/JPL/University of Arizona) Pg. 119: (main image) NASA-GSFC Scientific Visualization Studio; (inset) Natural color image from the Planetary Camera on the Hubble Space Telescope. (Hubble Space Telescope Comet Team /NASA) Pg. 120: Richard Turnridge Pg. 121: (top, left) Natural color image from the Wide Field/Planetary Camera 2 on the Hubble Space Telescope. (H. Hammel, MIT / NASA); (top, center) Brightness-enhanced image from the Long Range Reconnaissance Imager on the New Horizons probe. (NASA/Johns Hopkins University Applied Physics Laboratory/Southwest Research Institute); (top, right) Brightness-enhanced image from the Long Range Reconnaissance Imager on the New Horizons probe. (NASA/ Johns Hopkins University Applied Physics Laboratory/ Southwest Research Institute) Pg. 124: (top) Natural color image from the Solid State Imaging camera on the Galileo Orbiter. (Galileo Project, JPL, NASA); (bottom, left) Natural color images from the Solid State Imaging camera on the Galileo Orbiter. (Galileo Project, JPL, NASA); (bottom, right) Voyager 1 image taken from a distance of 490,000 km. (NASA/JPL/USGS) Pg. 125: (top) Natural color image from the Cassini-Huygens probe. (NASA/JPL/University of Arizona); (bottom) Natural color image from the Cassini-Huygens probe. (Cassini Imaging Team, Cassini Project, NASA) Pg. 126: (top, right) Image from the Stardust

Navigation Camera. (NASA/JPL-Caltech); (bottom) Natural color image from the Narrow Angle camera on Voyager 1. (NASA/JPL) Pg. 128: (top) Enhanced color image from the Solid State Imaging camera on the Galileo Orbiter, using violet, green and near-infrared filters. (NASA/JPL/University of Arizona); (bottom, left) False color image from the Narrow Angle camera on Voyager 2. (NASA/JPL); (bottom, right) Enhanced color image from the Solid State Imaging camera on the Galileo Orbiter, using violet, green and near-infrared filters. (NASA/JPL /University of Arizona) Pg. 132: (top) Natural color image from the Wide Field/Planetary Camera on the Hubble Space Telescope. (NASA/ESA/E. Karkoschka, University of Arizona); (bottom) Natural color image from the Solid State Imaging system on the Galileo Orbiter. (NASA/ JPL) Pg. 133: (top) Four-image mosaic from the Solid State Imaging system on the Galileo Orbiter. (NASA/JPL); (bottom) Image from the Solid State Imaging system on the Galileo Orbiter, with a spatial resolution (pixel size) of about 20 meters. (NASA/JPL/Brown University) Pg. 135: (top) Combination of color infrared data with a high resolution monochrome mosaic from the Solid State Imaging system on the Galileo Orbiter. (NASA/JPL/University of Arizona) Pg. 136: (top) Enhanced color image from the Solid State Imaging system on the Galileo Orbiter. (NASA/JPL/DLR); (bottom, left) Image from the Solid State Imaging system on the Galileo Orbiter. (NASA/JPL); (bottom, right) Image mosaic from the Narrow Angle vidicon camera on Voyager 1. (NASA/JPL) Pg. 137: Scaled mosaic of images from the Solid State Imaging system on the Galileo Orbiter. (NASA/JPL/Cornell University) Pg. 140: (bottom) 30-image natural color mosaic from the Wide Angle camera on the Cassini Orbiter. (NASA/ JPL/Space Science Institute) Pg. 141: (top) False color mosaic of 65 six-minute observations at infrared wavelengths from the Visual and Infrared Mapping Spectrometer on the Cassini Orbiter. (NASA/JPL/ASI/University of Arizona); (center, right) False color image from the Ultraviolet Imaging Spectrograph on the Cassini Orbiter. (NASA/JPL/University of Colorado); (bottom, left) False color infrared image from the Wide Angle camera on the Cassini Orbiter, using spectral filters at 752, 890 and 728 nanometers. (NASA/JPL/Space Science Institute); (bottom, center) False color infrared image from the Wide Angle camera on the Cassini Orbiter, using spectral filters at 752, 890 and 728 nanometers. (NASA/JPL/ Space Science Institute); (bottom, right) Artist's impression of the Cassini Orbiter, carrying the Huygens lander, both part of a cooperative mission by NASA, ESA and the Italian Space Agency. (NASA/JPL) Pg. 142: (top) Photograph of the London Underground logo courtesy of Wikimedia Commons; (center) Images form the Wide Field/Planetary Camera 2 on the Hubble Space Telescope. (NASA/The Hubble Heritage Team (STScI/AURA) / R.G. French, Wellesley College / J Cuzzi, NASA-Ames /L Dones, SwRI /J Lissauer, NASA-Ames); (bottom) Natural color image using red, green and blue filters from the Wide Angle camera on the Cassini Orbiter. (NASA /JPL/Space Science Institute) Pg. 143: (top) False color infrared image from the Wide Angle camera on the Cassini Orbiter. (NASA /JPL/Space Science Institute); (bottom, left) Photograph of British comedian Will Hay © Mirrorpix; (bottom, right) Natural color image from the Wide Field/Planetary Camera 2 (WFPC2) on the Hubble Space Telescope. (Reta Beebe, New Mexico State University/D.

Credits

Gilmore/L. Bergeron, STScI/NASA) Pg. 144: (top) Polar hexagon movies courtesy NASA/JPL/Space Science Institute; (center, left) Infrared image from the Narrow Angle camera on the Cassini Orbiter, using a spectral filter at 752 nanometers. (NASA/JPL/Space Science Institute); (center) Infrared image from the Narrow Angle camera on the Cassini Orbiter, combining polarized light at 746 and 938 nanometers. (NASA / JPL / Space Science Institute); (center, right) Northern hemisphere map showing Saturn's atmospheric temperature in the range -201 to -189 °C, from cold dark reds, to warmer bright orange and white, measured by the Composite Infrared Spectrometer on the Cassini Orbiter. (NASA /JPL/GSFC/ Oxford University); (Bottom) Near-infrared image using the 752 nm spectral filter on the Wide Angle camera of the Cassini Orbiter. (NASA/JPL/Space Science Institute) Pg. 145: Planetary Visions Pg. 148: Natural color mosaic of six images from the Narrow Angle camera on the Cassini Orbiter, covering a distance of 62,000 km (74,565 - 136,780 km from Saturn's center). (NASA / JPL / Space Science Institute); (top, left) False color infrared image at wavelengths of 1.0, 1.75 and 3.6 microns, from the Visual and Infrared Mapping Spectrometer on the Cassini Orbiter. (NASA /JPL /Space Science Institute); (top, right) Image from the Narrow Angle camera on Voyager 2. (NASA/JPL); (bottom, left) Natural color image from the Narrow Angle camera on the Cassini Orbiter. (NASA /JPL/Space Science Institute); (bottom, center) Natural color mosaic of 12 images from the Wide Angle camera on the Cassini Orbiter, taken over a period of 2.5 hours. (NASA /JPL/Space Science Institute); (bottom, right) Image from the Narrow Angle camera on the Cassini Orbiter. (NASA/JPL/Space Science Institute) Pg. 149: (top, left) Natural color image from the Narrow Angle camera on the Cassini Orbiter. (NASA /JPL/ Space Science Institute); (top, right) Image from the Narrow Angle camera on the Cassini Orbiter. (NASA/JPL/Space Science Institute); (bottom) Image from the Narrow Angle camera on the Cassini Orbiter. (NASA /JPL/Space Science Institute) Pg. 150: (top) NASA/JPL/Space Science Institute; (bottom) Color-coded optical depth map, derived from radio occultation observations at Ka-, X- and S-bands (094, 3.6, 13 cm wavelengths). Transmissions from Cassini's Radio Science Subsystem were recorded on Earth as the spacecraft passed behind the rings Pg. 151: Planetary Visions Pg. 153: Heikki Salo, University of Oulu, Finland Pg. 155: (bottom) Natural color view from the Wide Angle camera on the Cassini Orbiter. (NASA/ JPL/Space Science Institute) Pg. 156: (top) Simulated natural color image from the side-looking camera of the Descent Imager/Spectral Radiometer on the Huygens lander. (NASA/ JPL/ESA/University of Arizona); (bottom, right) Synthetic aperture radar image from the Radar Mapper on the Cassini Orbiter, operating at a frequency of 13.78 GHz. (NASA/JPL); (bottom, left) Artist's impression of the Huygens Lander on the surface of Titan. (ESA - C Carreau) Pg. 157: (top) False color synthetic aperture radar image from the Radar Mapper on the Cassini Orbiter, covering an area about 140 km across. (NASA/JPL) Pg. 158: (top) Natural color view from the Narrow Angle camera on the Cassini Orbiter. (NASA/JPL/Space Science Institute); (bottom, left) Natural color view from the Wide Angle camera on the Cassini Orbiter. (NASA/JPL/Space Science Institute); (bottom, center) False color view from the Wide Angle camera on the Cassini Orbiter, combining visible light (420 nm) with infrared (938 and 889 nm). (NASA/JPL/Space Science Institute); (bottom, right) Infrared view

from the Wide Angle camera on the Cassini Orbiter, using a filter at 938 nanometers. (NASA/JPL/Space Science Institute) Pg. 159: (top)Synthetic aperture radar image from the Radar Mapper on the Cassini Orbiter, operating at a frequency of 13.78 GHz. (NASA/JPL); (bottom) Simulated natural color image from the side-looking camera of the Descent Imager/Spectral Radiometer on the Huygens lander. (NASA/JPL/ESA/University of Arizona) Pg. 161: (bottom) Natural color image from the Wide Angle camera on the Cassini Orbiter. (NASA/JPL/Space Science Institute) Pg. 162: (top) Simulated perspective view based on detailed images from the Narrow Angle camera of the Cassini Orbiter, with a height exaggeration of about 10 times. (NASA/JPL/Space Science Institute/Universities Space Research Association/Lunar & Planetary Institute); (bottom, left) Mosaic of two images from the Narrow Angle camera on the Cassini Orbiter. (NASA/ JPL/Space Science Institute); (bottom, right) Visible light image from the Narrow Angle camera on the Cassini Orbiter. (NASA/JPL/Space Science Institute) Pg. 163: (left) Enhanced color image mosaic from the Narrow Angle camera on the Cassini Orbiter, using infrared, green and ultraviolet filters. (NASA/JPL/Space Science Institute); (right) Enhanced color image mosaic from the Narrow Angle camera on the Cassini Orbiter, using infrared, green and ultraviolet filters. (NASA/JPL/Space Science Institute) Pg. 166: (top) Enhanced color image mosaic from the Narrow Angle Camera on the Cassini Orbiter, combining detailed images using the clear filter with color images using infrared, green and ultraviolet filters at 752, 568 and 388 nm. (NASA/JPL/Space Science Institute); (bottom, left) Image from the Narrow Angle Camera on the Cassini Orbiter. (NASA/JPL/Space Science Institute); (bottom, right) Mosaic of two clear-filter images from the Narrow Angle Camera on the Cassini Orbiter, showing features as small as 36 meters across. (NASA/JPL/Space Science Institute) Pg. 167: Image from the Narrow Angle Camera on the Cassini Orbiter. (NASA/JPL/Space Science Institute) Pg. 170: (main image) Image from the Narrow Angle camera on the Cassini Orbiter. (NASA/JPL/Space Science Institute); (left) Natural color image from the Narrow Angle camera on the Cassini Orbiter. (NASA/JPL/Space Science Institute); (center) Temperature map based on data from the Composite Infrared Spectrometer on the Cassini Orbiter. Temperature ranges from -196 Celsius (blue) to -181 Celsius (yellow). (NASA/JPL/GSFC/SWRI/SSI); (bottom) Enhanced color mosaic of images taken with ultraviolet, green and infrared filters, combined with a detailed image taken through the clear filter, from the Narrow Angle camera on the Cassini Orbiter. (NASA/JPL/Space Science Institute) Pg. 174: (top) Extreme color enhanced image using infrared, green and ultraviolet filters of the Narrow Angle camera on the Cassini Orbiter. (NASA/ JPL/Space Science Institute); (bottom) NASA/JPL/Space Science Institute Pg. 178: (bottom, left) Natural color image from Voyager 2 Narrow Angle camera using blue, green and orange filters. (NASA/JPL); (bottom, center) Voyager 2 Narrow Angle camera image using blue, green and orange filters. (NASA/JPL/USGS); (bottom, right) False color image from Voyager 2 Narrow Angle camera using ultraviolet, violet and orange filters. (NASA/JPL) Pg. 179: (top) Image from the Hubble Space Telescope Wide Field/Planetary Camera. (NASA/ESA/M. Showalter, SETI Institute); (bottom, left) Composite image from the Hubble Space Telescope's Wide Field/Planetary Camera. (NASA/ESA/M. Showalter, SETI

Institute/Z. Levay, STScI); (bottom, center) 15-second exposure through the clear filter on Voyager 2's Narrow Angle camera. (NASA/JPL); (bottom, right) 96-second exposure through the clear filter on Voyager 2's Wide Angle camera. (NASA/JPL) Pg. 180: (top) Image from the Hubble Space Telescope Wide Field/Planetary Camera. (NASA/ESA/L. Sromovsky and P. Fry, University of Wisconsin / H. Hammel, Space Science Institute/K. Rages, SETI Institute); (left) Image of an oil painting of German-born British astronomer Sir Frederick William Herschel by John Russell RA courtesy of Wikimedia Commons Pg. 181: (top) Image from the Hubble Space Telescope Wide Field/Planetary Camera. (NASA/ ESA/M. Showalter, SETI Institute); (center) False color image taken with the 10-meter Keck 2 telescope's Near Infrared Camera. (W M Keck Observatory/Larry Sromovsky, University of Wisconsin); (bottom) A composite of images from the 10-meter Keck 2 telescope, using H-band and K-band filters. (W M Keck Observatory/Marcos van Dam) Pg. 184: False color image using 1.2 and 1.6 micron wavelengths from the NAOS-CONICA infrared camera on the European Southern Observatory's 8.2-meter Very Large Telescope (VLT), Paranal, Chile. (ESO) Pg. 185: (left) Near infrared image in the 2.2 micron Ks-band from the ISAAC multi-mode instrument on the 8.2-meter Very Large Telescope at the European Southern Observatory, Paranal, Chile. (ESO) Pg. 186: (bottom, left) False color image through green, violet and ultraviolet filters on Voyager 2's Narrow Angle camera. (NASA/JPL); (bottom, right) NASA/ESA/ L. Sromovsky, University of Wisconsin, Madison /H. Hammel, Space Science Institute/K. Rages, SETI Pg. 187: (top) False color image taken with the Near Infrared Camera and Multi-Object Spectrometer (NICMOS) on the Hubble Space Telescope. (NASA/JPL/STScI) Pg. 190: (bottom) Two images from Voyager 2's Wide Angle vidicon camera using the clear filter, at a 590 second exposure. (NASA/JPL) Pg. 191: (bottom) Voyager 2 Narrow Angle vidicon camera image. (NASA/JPL) Pg. 192: Voyager 2 Narrow Angle vidicon camera image using violet and orange filters. (NASA/JPL) Pg. 193: (top, left) Voyager 2 Narrow Angle vidicon camera image using green and clear filters. (NASA/JPL); (top, right) Mosaic of five images from Voyager 2's Narrow Angle vidicon camera image using clear, orange and green filters. (The Voyager Project, NASA); (bottom) False color image from the Hubble Space Telescope (HST) Wide Field/Planetary Camera (WFPC2). (NASA/JPL/STScI) Pg. 195: Mosaic of images, using orange, violet and ultraviolet filters, from Voyager 2. (NASA/JPL) Pg. 196: Planetary Visions Pg. 198: (top) Artist's impression. (NASA/Planetary Visions); (center, right) Joe Zeff Design; (bottom, left) Artist's impression. (NASA / ESA / G. Bacon, STScI)) Pg. 199: (center, left) Composite of 16 exposures from the Advanced Camera for Surveys on the Hubble Space Telescope. (NASA/M. Brown, Caltech); (center, right) Sum of 16 exposures from the Advanced Camera for Surveys on the Hubble Space Telescope. (NASA/M. Brown, Caltech); (bottom) Photograph of Dutch-American astronomer Gerrit Pieter Kuiper ©Dr. Dale P. Cruikshank Pg. 200: (bottom, left) (NASA-JHUAPL-SwRI) (bottom, right) Ground-based image from the Canada-France-Hawaii telescope in Hawaii. Space telescope image from the Faint Object Camera on the Hubble Space Telescope, taken in 1990. (NASA/ESA) Pg. 201: (bottom) (NASA/Johns Hopkins University Applied Physics Laboratory/Southwest Research Institute) Hubble

Space Telescope. (Dr R Albrecht, ESA/ESO Space Telescope European Coordinating Facility/NASA) Pg. 202: (top) Simulated view based on a global map of estimated true color, derived from multiple observations from the Hubble Space Telescope. (Eliot Young, SwRI, et al/NASA); (bottom) Artists impression of the surface of Pluto. (ESO/L Calçada) Pg. 203: (top, left) (NASA-JHUAPL-SwRI) (top, right) Image from the Advanced Camera for Surveys on the Hubble Space Telescope. (NASA/ESA/H Weaver, JHUAPL/A Stern, SwRI/ HST Pluto Companion Search Team); (bottom) Image form the Advanced Camera for Surveys on the Hubble Space Telescope. (H Weaver, JHU-APL/A. Stern, SwRI/ HST Pluto Companion Search Team/ESA/NASA) and (NASA/APL/SwRI). Pg. 204: Image from the Hubble Space Telescope's Advanced Camera for Surveys. (NASA/ESA/M. Brown, California Institute of Technology) Pg. 205: Touch Press Pg. 206: (left) Artist's impression. (NASA/ESA/A. Field, STScI); (right) Image from the Hubble Space Telescope's Advanced Camera for Surveys. (NASA/ESA/M. Brown, California Institute of Technology) Pg. 207: (left) Artist's impression. (NASA/ESA/A. Field, STScI); (right) Image from the 10 meter-diameter Keck II telescope, using the Keck Observatory Laser Guide Star Adaptive Optics system in 2005. (NASA/M Brown)

Pg. 208: Joe Zeff Design Pg. 209: Photograph of Dutch astronomer Jan Hendrik Oort ©Leiden Observatory and Wikimedia Commons Pg. 210: (top) Photograph taken just after sunset from the European Southern Observatory at Paranal in Chile. (S. Deiries/ESO); (bottom) Plate photograph taken on 29th May 1910, published in the New York Times on 3rd July 1910. (Yerkes Observatory, University of Chicago) Pg. 212: (top, left) Mosaic of images from the Halley Multicolor Camera on the Giotto probe. (ESA/MPAe, Lindau); (top, right) Mosaic of 68 images from the Halley Multicolor Camera on the Giotto probe. (ESA/MPAe, Lindau); (bottom) Image of a section of the Bayeux Tapestry showing Halley's commet ©Reading Museum (Reading Borough Council). All rights reserved. Pg. 213: (top) Photograph taken on March 8, 1986 from Easter Island. (NASA-NSSDC/W. Liller); (bottom, left) Photograph using a telephoto lens. (ESO); (bottom, center) Microscope photograph. (NASA-JSC); (bottom, right) Magnified photograph. (NASA-JSC) Pg. 214: Image from the Advanced Camera for Surveys/Wide Field Camera on the Hubble Space Telescope. (NASA/ESA/H. Weaver, JHU-APL/M. Mutchler, Z. Levay, STScI) Pg. 215: Photograph of trees damaged in the Tunguska explosion courtesy of Wikimedia Commons. Pg. 216: (left) TA Rector/NOAO/AURA/ NSF; (right, top) Image from the Stardust Navigation Camera (NASA/JPL-Caltech); (right, bottom) Infrared image, at a wavelength of 24 microns, from the Multiband Imaging Photometer on the Spitzer Space Telescope (NASA/JPL-Caltech/ W Reach, SSC-Caltech) Pg. 217: (top, left) Image from the Medium Resolution Instrument on the Deep Impact probe. (NASA/JPL-Caltech/UMD); (top, right) Image from the High Resolution Instrument on the Deep Impact fly-by spacecraft. (NASA/JPL-Caltech/UMD); (bottom) Composite of 15 5-minute exposures from a Canon EOS 350D through a 130mm refracting telescope. (Ivan Eder)

Index

Index